THE EARTH
IN THE
LOOKING
GLASS

THE EARTH IN THE LOOKING GLASS

Lloyd Darden

Anchor Press / Doubleday
Garden City, New York
1974

Library of Congress Cataloging in Publication Data

Darden, Lloyd.
The earth in the looking glass.

Includes bibliographical references.
1. Geography. 2. Remote sensing systems.
I. Title.
G70.4.D37 910
ISBN 0-385-02595-5
Library of Congress Catalog Card Number 73-9151

Printed in the United States of America
First Edition

To Dotty

Acknowledgments

It was Dr. Story Musgrave who jolted me into a genuine awareness of what satellites might help man to achieve for himself on earth. During our chance encounter in February 1971, in Washington, D.C., Musgrave (chemist, medical doctor, statistician, and later a backup astronaut for Skylab) revealed to me some of the enormous benefits man could attain by studying his own planet through the eyes of a satellite. Someone should write a book on the subject, we agreed, and as I later became more and more excited by "earth observations," I decided to nominate myself for the task.

In the two and a half years that followed, my path crossed those of more than a hundred scientists—each of them involved in a vital matter such as searching for energy or food—and all of them observing the earth in pictures and data collected by satellites. Two scientists, however, must bear a particular burden of responsibility for whatever these pages hold. They lent their time, insight, and good humor to my continuing education in the field of studying the earth from space. They are Anthony J. Calio, physicist and Director of Science and Applications at Johnson Space Center, Houston, and Ira Bechtold, my neighbor in La Habra, California, and one of the pioneers in the use of satellite information for geology. Both Tony and Ira operate with horizons which dramatically exceed their respective scientific disciplines; their contributions, much like the view from a satellite, have been broad indeed.

Among the many other sources to whom I am indebted, I alphabetically list the following:

Dr. Monem Abdel-Gawad, Rockwell International Science Center; Robert Aldrich, U. S. Forest Service; Howard Allaway, National Aeronautics and Space Administration; George Armstrong, State of California; John Arveson, Ames Research Center; Tom Beall, Cattle Fax; Dr. Joseph Behar, California Air Pollution Research Center; Andrew Benson, University of

California, Berkeley; Dr. Martin Biles, Atomic Energy Commission; Dr. Leonard Bowden, University of California, Riverside; Dr. William Campbell, U. S. Geological Survey; Dr. Paul Carlson, U. S. Geological Survey; Dr. David Carneggie, University of California, Berkeley; Scott Carpenter, astronaut; Dr. Walter Clark, Eastman Kodak (retired); Walter Cleveland, McDonnell Douglas Company; Virginia Coleman, University of California, Riverside; Dr. Robert Colwell, University of California, Berkeley; Richard C. Culler, Department of the Interior, Water Resources Division; Earl Davis, State of California; George DeKoven, Hughes Aircraft Company; John Donnelly, National Aeronautics and Space Administration; Vernon Dvorak, National Environmental Satellite Service; Dr. J. T. Eaton, National Center for Earthquake Research; Edward Ferguson, National Environmental Satellite Service; Robert Fett, National Environmental Satellite Service; Dr. Charles Frazee, State University of South Dakota; Dr. Per Gloersen, National Aeronautics and Space Administration; Hermie Gloria, Ames Research Center; George Gryc, U. S. Geological Survey; William Harting, Tri-State Regional Planning Commission, New York; Robert Heller, U. S. Forest Service; Ward Henderson, U. S. Department of Agriculture; Edward Hengen, John Deere Harvester Works; Lester Hubert, National Environmental Satellite Service; James Huning, University of California, Riverside; Claude Johnson, University of California, Riverside; Pat Jones, National Aeronautics and Space Administration; Ernest Lathram, U. S. Geological Survey; Dr. Dale Leipper, U. S. Naval Postgraduate School; Mark Liggett, Argus Exploration Company; Dr. A. O. Lind, University of Vermont; Dr. Paul Lowman, National Aeronautics and Space Administration. Dr. Norman Macleod, American University (then of NASA); Charles Matthews, National Aeronautics and Space Administration; Dr. Paul Maughan, Earth Satellite Corporation; Dr. Robert McDonald, National Aeronautics and Space Administration; Joe McRoberts, National Aeronautics and Space Administration; John Mehos, Liberty Fish and Oyster Company; Dr. Mark Meier, U. S. Geological Survey; Dr. Paul Merifield, University of California, Los Angeles; John Millard, Ames Research Center; Richard Mittauer, National Aeronautics and Space Administration; Harvey K. Nelson, Bureau of Sport Fishing and Wildlife; William Nordberg, National Aeronautics and Space Administration; Allen Pearson, National Severe Storms Forecast Center; Dr. Wayne Pettyjohn, Ohio State University; Dr. James N. Pitts, Jr., California Air Pollution Research Center; Dr. John Place, U. S. Geological Survey; Walter Planet, National Environmental Satellite Service; Dr. C. C. Reeves, Texas Tech University; Dr. Ronald Reinisch, Ames Research Center; Dr. R. J. Renard, U. S. Naval Postgraduate School; Dr. Ernest Rogers, Aerospace Corporation; D. S. Ross, International Imaging Systems; Dr. Jane Schubert, National Aeronautics and Space Administration; Jack Sherman, National Oceanic and Atmospheric Administration; Dr. Joel Silverstein, Rockwell International Science Center; Dr. Merritt Stevenson, Inter-American Tuna Commission; Dr. Robert E. Stevenson, Office of Naval Research; Dr. Gordon Thayer, National Oceanic and Atmospheric Administration; Dr. Eugene Thorley, University of California, Berkeley; Andrew Timchalk, National Environmental Satellite Service; Brian Topolovski, U. S. Army Corps of Engineers; Richard Underwood, National Aeronautics and Space Administration; A. H. Waite, inventor; Douglas Ward, National Aeronautics and Space Administration; Dr. Peter Ward, National Center for Earthquake Research; Dr. Ellen Weaver, Ames Research Center; Dr. Edgar Weyburn, Sierra Club; Dr. Charles Wier, University of Indiana; Harold Woolf, National Environmental Satellite Service.

Apart from this informational assistance were other forms of help, such as artwork from CNP Litho Company, Inc., and stimulation from a trio of editorial people: Luther Nichols, Doubleday's West Coast editor, and Maureen Mahon, science editor (located in New York), both provided guidance for which I am most grateful. Ironically, I find that the last person on this list is the one who gave me my first encouragement—Don James, one of Portland, Oregon's, much-published authors of books.

L.D.

CONTENTS

Part I

THE OVERVIEW

1. The Looking Glass

"The earth is not only shrinking. It is also getting *larger*."

Charles Matthews paused after dropping this paradoxical comment on the press corps assembled in the Los Angeles Biltmore Hotel. I utilized the momentary lull to turn around in my chair and glance at the faces in the rows behind me. As an independent writer and not a member of the press *per se*, I was interested in seeing how the media people were reacting to Matthews. Their raised eyebrows and half-smiles seemed to express a "Nice, but what's this all about?" sort of response. Matthews' remark was hardly what they expected from the National Aeronautics and Space Administration; NASA people were supposed to talk in positive terms about liftoffs and interplanetary probes—not in obscure riddles about the earth.

From previous conversations with Matthews, I happened to know what he had in mind with his words about the earth "getting larger," and it held a meaning that would become clearer to me in 1973 and 1974 when the controversial energy shortage would erupt. But on the day of this press conference, July 19, 1972, the speaker's enigmatic comment hinted of more drama than newsmen were ready to understand or accept. Although Matthews held the new and expansive title of "Director of Applications" at NASA, he was an unknown to most of the reporters. Furthermore, space activities had become a fading news topic at best, and reporters considered the subject of Matthews' conference to be particularly low on the NASA priority pole; the press event of this day was a scientific experiment—a research project, if you will—with none of the old razzle-dazzle of the space world. Unobtrusively, it would have its beginning four days later on July 23, near the rather remote coastal town of Lompoc, California. And regardless of any esoteric comments which Charles Matthews might make today at the Biltmore, neither the reporters nor anyone else outside the scientific community was likely to become terribly excited about it.

Yet Lompoc on July 23 was providing a historical moment which, in the minds of some scientists, was to be as significant as the day when the atom had been split. Lompoc was an event designed to trigger a flow of totally new information into planet earth—unique information which related only to the earth, yet which was collected entirely in space. Information which could allow man, for the very first time in all history, literally to begin to manage his own planet.

The Lompoc event which was to lead man to this new level of competence was the launching of a single, unmanned space capsule. It would be known as the first Earth Resources Technology Satellite (ERTS-1). ERTS was to be a new kind of space satellite (a robot, some called it) with an almost inconceivable assignment; it was being sent 570 miles into the sky, whence it was to detect a multitude of things existing in the world beneath our feet—things which man had looked at, tripped over, or even slept on for the last several millennia, without ever recognizing.

However, the full potential of ERTS was somehow not surfacing at the press conference. Reporters were jotting down the scientific concepts being presented to them; but it was obvious from their questions that they were not really grasping the tremendous impact the project might have on their lives. Yet if these same newsmen could have somehow been spirited out of the Biltmore and inconspicuously dropped into any of several major oil companies nearby, their conception of ERTS would have changed swiftly and markedly.

The Energy Probe

Had such a hypothetical visit taken them to one particular oil company, they would have been exposed to a certain tall, sandy-haired, middle-aged research geologist who was much more keenly interested in earth observations than he chose to discuss publicly. His job was to find better ways of locating oil; for that reason, he had been closely watching the development of ERTS for years and had become intimately familiar with it. He knew, for example, that ERTS would be taking a picture of the earth every twenty-five seconds.

"Each of those pictures," he later commented to me, "covers an area 115 by 115 miles. Now in this company, we are presently using reconnaissance airplanes to find areas of probable oil deposits. It takes *thousands* of aerial pictures to cover a 115-mile square, and it costs us more than $100,000." At that rate, it would require billions of dollars to cover the world with airplanes even once, and as a result, the petroleum search had been limited to scattered portions of the globe. But ERTS, sending back the equivalent of a $100,000 picture every twenty-five seconds, had the potential to make global search suddenly feasible.

The oil of the world, from the Middle East to Venezuela to Alaska's Prudhoe Bay, obviously had been discovered quite satisfactorily up until now without benefit of satellites. But search time was shorter now. Despite a widespread feeling that the energy crisis of the mid-70s was "contrived," there were few who doubted the reality of a longer range shortage; therefore, oil and gas strikes which would not normally have occurred until several decades hence were being urgently sought to fill today's energy gap (while future forms of fuel were being developed).

ERTS, as it turned out, would respond to help fill the gap more quickly, quietly and yet more spectacularly than anyone had imagined possible; just four days after its launch, ERTS would obtain a remarkable picture of the Alaskan north slope. It would be a picture which the following year would lead oil companies to a second great north slope exploration area—an area exhibiting signs of becoming another Alaskan oil bonanza, conceivably surpassing the Prudhoe Bay fields that had suggested the Alaskan pipeline. ERTS may have pointed out a major fuel source for the 1980s.

It was this worldwide search potential which was attracting a cross section of scientists to ERTS—not only to seek out oil but to find other wellsprings of fuel and energy and to find dramatic new ways of controlling our planet's air and water . . . and, in what might be the most immediately promising concept of all, to find new sources of food for the world.

Was all this possible? Could this "eye" in the sky, looking at the earth nearly 600 miles away, actually find resources and opportunities unknown to the earth's inhabitants themselves? Could a satellite serve the earth as a conveniently ubiquitous watchdog? These were questions which in some cases could be answered immediately and affirmatively—and in others only gradually and complexly over a period of months and years, as ERTS findings were collected and interpreted and utilized.

In the meantime, ERTS was to be observed this very week, not yet in the name of science, but simply as another example of aerospace engineering. For before any contributions to science could begin, NASA must first put the satellite into the sky.

Since it was to be launched southward, to initiate an orbit that would regularly shuttle it between Arctic and Antarctic, it required a liftoff area that was completely free of population to the south. Conveniently jutting out from the U.S. Pacific coastline was a finger of land, once best known as the launch site for Atlas war missiles, but now more famous as a bucolic center for the nation's flower-growing industry. This was Lompoc, California.

The ERTS Launch

Three days after the press conference, I was on my way to the launch site, driving at night from my home in La Habra, California, northward up the coast freeway. Traveling without a motel reservation, and having some idea about the size of Lompoc, I was a bit uneasy. While I didn't expect the area to be blossoming overnight into a Cape Kennedy, where a million people had customarily thronged to the Apollo launches, I nevertheless imagined that the small town would be bulging at the seams on this eve of the launch.

Because of my concern, I turned off the freeway at Santa Barbara, some thirty miles before reaching the Lompoc cutoff, and stopped at a motel.

"You're lucky to get this room," the night clerk said, smiling. "My last one. Town's full tonight."

"Really?" I was pleasantly surprised to think that the Lompoc overflow had reached this far. "ERTS?"

The night clerk appeared baffled by my monosyllabic grunt. He apparently thought I had gas pains. Deciding to ignore them, he said, "Guess you're in town for the national horse show, like everyone else?"

I winced at that, signed the guest register, and left for my room.

The next morning I turned my car away from what I presumed could only be horse-show traffic and continued on to Lompoc. There, along with a mere five hundred invited guests, a smattering of the press, and precious little ceremony, I watched as the potentially most productive scientific effort in history was lifted rather anonymously into space.

In terms of thrust, the blastoff was smaller than Apollo launches, but a combination of liquid and solid fuel systems provided ERTS with a double-header in fireworks. It was, in a sense, more spectacular than the Cape Kennedy launches, and yet it would cause barely a ripple in the news media.

The significance of ERTS, of course, was yet to come, the launch being only a kind of symbolic golden spike. NASA people might have liked the symbol to have been more widely recognized, of course, but they were not terribly surprised that it had not been. ERTS was, after all, carrying cargo of impersonal scientific instruments instead of a crew of flesh-and-blood astronauts, and, moreover, it had come a decade too late for the great accolades of spacedom.

"Things were different," aerospace people were fond of saying, "back in the sixties." And there was no doubt about that. The 1960s had been the golden era of space exploration. It had been a period like nothing in history before—a decade of "wows"—and everyone was caught up in it. It had begun in 1961 with Alan Shepard, the first American in space, the A-OK hero, and it had climaxed in 1969 with Neil Armstrong and the Giant Step for Mankind. It had been the decade of the astronaut and the space engineer. The space walk. The flag on the moon. Man's rendezvous with interplanetary destiny. Even the most blasé among us were incredibly excited.

And then, kapow, it was over. Like a clap of thunder. Excitement had dwindled and money was cut. The Apollo program was truncated with three missions to go. Space engineers were suddenly unemployed. Alan Shepard wound up with an early case of "future shock"; for every person who still remembered him as a hero, another saw him as a heavy who squandered our money hitting a golf ball on the moon.

But all during the time that the popularity of lunar flight was waning, a completely separate low-budget project was silently moving ahead. It had begun in the mid-sixties, halfway between Shepard's cheery A-OK flight and his unpopular golf stroke, and it had started quite by accident, with NASA virtually stumbling onto a great new opportunity. A funny thing had happened on the way to the moon; *someone had looked back at the earth.*

Quite spontaneously, that had been the beginning of the era now about to become so significant. The view of the earth from space had proved exhilarating, not only for the astronauts but for all those who looked at the pictures they brought back from their missions. Several happenstance photographs of the earth had stimulated scientists to begin studying satellite pictures, casually at first, until they realized how much earth detail they could distinguish. Meteorologists had already evidenced strong interest in satellite pictures. Now geologists did the same, and one by one the scientific disciplines—agriculture, forestry, oceanography—embraced the knowledge of this new medium and wedded themselves to it. The casual study of pictures gave way to more serious projects which eventually coalesced into the plans for ERTS and Skylab, two satellites designed to introduce a new technology known as "remote sensing." (For the moment, let us content ourselves by thinking of remote sensing as simply the taking of pictures of the earth from space.) By the time the 1960s and the golden era of space engineering were drawing to a close, scientists were calling for a new decade in space, a *science* decade, a period in which they could comprehensively study their world from the outside. They wanted to examine and study their own planet in a way never possible before, seeing it not only as it appeared now but also as it would evolve before their eyes throughout the 1970s.

The Planet Shudders

On December 23, 1972, at 12:30 A.M., an earthquake virtually turned Managua, Nicaragua, upside down, killing thousands of people and leveling the homes of more than a quarter million. Before the ashes and pungent dust had settled the following day, the eyes of satellite ERTS, making its scheduled pass over Central America, were trained on Managua. ERTS was soon followed by a NASA aircraft, using ERTS-type remote-sensing gear to observe the city from low and medium altitudes. The pictures and data were brought back to Houston, to the White House, and to the National Center for Earthquake Research in Menlo Park, California.

The results were revealing, to say the least. Whereas an

averaging of seismographs around the hemisphere on December 23 had indicated only a 5.6 Richter magnitude, the NASA pictures indicated that the tremor, with its epicenter in the congested downtown of Managua, had been one of the most damaging quakes of the century.

In California, where virtually everyone is earthquake-conscious, newsmen began to descend on the Menlo Park center, seeking comments; they were not disappointed. Just a week after the NASA sensing had taken place, the center's chief for earthquake research, Dr. J. T. Eaton, made a startling statement on national television: "Earthquake forecasts will in ten years be as reliable and precise as hurricane warnings are today." (The eruption of volcanoes might be equally predictable, he had commented to me earlier the same week.) His optimism, of course, was electrifying in the more earthquake-prone communities of the United States.

While Eaton was making dramatic prognostications in California, a group of other geophysicists from his center was already in Managua, laying out a network of seismic recording devices. Along with these recorders placed around the city itself, an extensive observation of nearby San Cristóbal volcano was about to be undertaken by ERTS in cooperation with the Nicaraguan government. The scientists suspected that San Cristóbal might erupt in 1973, and they were counting on ERTS to provide an advance alarm and preclude another mass disaster.

Protection from disaster, however, was just one of the objectives of the monitoring. From a research standpoint, the Managua tragedy had been convenient. For more than a year before the earthquake, Menlo Park geophysicists had been working on a plan to use ERTS to monitor San Cristóbal and a score of other Central and North American volcano areas. Thereafter, as Managua trembled with aftershocks throughout 1973, it provided students of earthquakes and volcanoes with the best laboratory the world could offer. ERTS would receive daily transmissions from Nicaragua and the other areas in the experimental volcano chain, transmissions that would describe relevant underground activity and contribute valuable data toward the understanding of all volcanoes. If Eaton's optimism over forecasting both volcanoes and earthquakes was justified, much of its basis lay in this and other projects being conducted by satellite.

To Explore or to Survive

Earthquakes and volcanoes are only one small part of the incredible program known as ERTS. Hundreds of diverse projects began utilizing the satellite in 1972. Yet in the year that followed I was continually struck by the fact that virtually no one outside the world of science was aware that man had begun meticulously scrutinizing his own planet from space. And no small wonder. The scientists, who did realize what was occurring, had characteristically maintained discreet silence about their respective sensing projects and their high hopes of important discoveries over the next several years. Scientists are, quite justifiably, wary of oversimplified reporting which causes the public to expect too much of them, too soon. (Eaton, with his startling promise about earthquake predictions, was being unusually frank.)

Among nonscientists, even those persons whose jobs or avocations seemed relevant to ERTS, the observing of earth from satellites is today still almost unknown. Most of the half million persons employed in California's aerospace industry have scarcely heard of remote sensing. Furthermore, the nation's multitudes of self-styled environmentalists, whose causes stand much to gain from a genuine knowledge of the earth, are equally unenlightened. Remote sensing has truly matured into what seems to be one of the world's best kept secrets since the invasion plans for D-Day.

This utter lack of awareness of sensing has allowed space-oriented people and environmentalists oftentimes to look on themselves as polarized factions. In the very early 1970s, environmentalists were in constant skirmishes with aerospace advocates (just as they are with the power companies and the energy seekers in the mid-1970s). The aerospace-environmentalist debate had by 1972 achieved classical status, and in that period you could hear it being waged every day, almost anywhere in the United States. Even today, it occurs frequently.

The typical argument commences with some space enthusiast relating his theories on interplanetary adventure. "The rest of this century," the reasoning begins, "offers an opportunity for us to press on in space, to probe the unknown. To conquer the universe. Because it is there."

Whenever such a case is being presented, any environmentalist present usually squirms impatiently, the way a person does when he thinks he can rip an argument to shreds. "You're an explorer by inclination," the environmentalist begins, when he eventually has his chance to talk. "Good enough. On the other hand, I'm a *survivalist,* and I'm for dropping this space foolishness and knuckling under to the problems here on the good earth. I just read the report by the Club of Rome."[1] (And that, he comments, has to be one of the best-balanced brain trusts ever to come down the pike.) "Those people have dissected the urgent problems we face, and they've extended them into the future, on graphs. Population against food. Industrialization against pollution. Energy use against fuel depletion. What they've projected from all this is a nightmare! We're running out of everything but people and smog." At this point the survivalist usually concludes his argument by pointing downward at the "good earth," saying something like "*This* is the planet we need to be conquering. *This* is where our dollars should be spent."

"But why pick on the space program for your money?" the explorer then asks in exasperation. "We're already down nearly 40 per cent from our peak years."

"And high time," the survivalist typically responds. "For ten years, NASA had more wealth than an Arab sheikdom."

The explorer shakes his head rather painfully at this point. "Actually, we averaged only a little more than 2 per cent of the national budget throughout the 1960s."

(The trouble with the 1970s is that people keep comparing

them with the 1960s. And yet, from the scientist's viewpoint, this newer decade is undergoing more changes and approaching more crossroads than had ever been imagined in the most soaring moments of the sixties.)

The Scope of the 1970s

ERTS was not to be assigned the task of monitoring the 1970s alone. The satellite Skylab had, after a shaky beginning, joined ERTS in early 1973. Trouble-plagued as it was, Skylab's beginning was more auspicious than ERTS's, and the media had given the manned satellite ample attention beginning well in advance of the launch. On May 10, 1973, just a few days before the initial Skylab liftoff, CBS telecaster Walter Cronkite philosophized to his evening news audience that "America's space program is turning its eyes back to earth."

If Cronkite's observation was apt, it was also nine months late. Like the rest of the media people, Cronkite seemed unaware that ERTS had turned its eyes earthward the summer before, with a more comprehensive view of the world than Skylab was ever intended to have. Skylab was only the tip of the remote-sensing iceberg.*

During the months that newsmen were reporting on Skylab, there were other science-related news matters considerably more prominent in the minds of the American public—the energy shortage and the grain and meat scarcities. These matters were among the five elements the Club of Rome a year before had underscored as critical for global survival: food, energy and natural resources, industrialization, pollution, and population. Less well known was the fact that Skylab, and ERTS even more so, was making progress in four of these five vital areas (all except population control).

Within the first months of ERTS, examples of solid progress

* Skylab, however, unlike ERTS, had several missions other than remote sensing: medical and bioscience experiments, studies of the effects of space flight, investigations of the sun and its effects on the earth, and experiments in weightless technology.

by remote sensing occurred in each of these four areas. First, the matter of food sources. Satellites have been used to seek out potentially valuable farmland in areas which were always previously considered uncultivable. By 1972, some such land was already being recognized in ERTS pictures of a South Dakota Sioux Indian reservation—ironically, just a few miles from Wounded Knee, where the Sioux in 1973 staged a poverty demonstration. In areas which Indians and others had previously rejected as worthless and barren, there were extensive fields now perceived to have a high potential for growing grain.

Second, natural resources. While satellites were leading oil companies to vast new petroleum areas yet to be developed in North America, a leading consulting geologist, Ira Bechtold, studied an ERTS picture of Nevada with a quite different purpose. Small, circular patterns which he saw in the picture indicated to him probable locations of geothermal power sources. Bechtold saw the circular patterns repeated in picture after picture, and his client, an exploration company interested in geothermal development throughout the western states, was quietly pursuing Bechtold's findings.

Third, the needs of an industrial world. In such areas as New York, New Jersey, and Connecticut, future power-plant sites capable of satisfying the demands of both power company and environmentally motivated community are as difficult to find as they are urgently needed. Information from ERTS is already being used to seek out unique plant locations—locations that may not only help the power companies avoid brownouts but also allay the residents' fears of thermal and nuclear contamination.

Fourth, the detection of pollution and general management of the earth's environment. Within days of the Lompoc launch, pictures significant to environmental matters were being telemetered from ERTS to earth. One picture showed an S-shaped line in the ocean off the New Jersey shore. Investigation proved the "S" to have been the acid wake of a zigzagging New York garbage ship, dumping refuse too close to the shore. The other ERTS picture dramatically revealed a forest fire burning in

Alaska, in a remote area where fires were seldom detected by conventional means. Both pictures are dramatic testimonials for the environmental value in a satellite's comprehensive view of the earth.

These random examples are representative of hundreds of vital projects. Streams of data are being collected from all over the globe by ERTS, then transmitted back to one of the three ERTS receiving stations on earth,† and finally distributed worldwide to the sponsoring scientists. Hundreds of millions of bits of information are being collected every day and fed in to 328 scientific projects.

ERTS offers a means of monitoring the earth so that man can make his planet more productive, more responsive to the increasing needs of its inhabitants. The words of Charles Matthews are apt indeed: We are trying to make the earth grow rather than shrink.

The Looking Glass

"ERTS offers a whole new kind of exploration," geologist Ira Bechtold has said. "Almost no one realized that until ERTS came along, NASA had taken more pictures of the moon than of our own world! But now," says Bechtold, "ERTS is really serving as a mirror for the earth. It's allowing us to see things all around ourselves which we could never have found simply by exploring here on the ground. Up till this moment in history, we've been like a horde of ants wandering around on the cheek of the Mona Lisa."

Not so any more, now that ERTS is in orbit. ERTS is reflecting the entire canvas of the world for us to see. While we may stand antlike on the earth, we can now look out at the mirror and, for the first time, see reflections of large segments of the total picture around us.

The Mona Lisa analogy is subscribed to by a number of

† Goddard Space Flight Center, Greenbelt, Maryland, is the principal receiving station, as well as ERTS's control headquarters.

prestigious scientists—among them conservationist Dr. Robert Colwell of the University of California at Berkeley—although the concept is still held suspect by some lay survivalists. One such survivalist once asked Colwell, "Don't we really need a *close-up* view of things, rather than a far-out view from space? Isn't the earth on a dizzying spin toward calamity? . . ."

"With time bombs ticking all over the place. Agreed," Colwell answered (being a survivalist of no small repute himself). "But," he added, "we have to *find* them before we can disarm them."

Finding the problems of the world, explained Colwell, is the crux of the task ERTS is performing. California, which lays claim to perhaps the world's greatest assemblage of renowned scientists (and an almost equal concentration of environmental problems), has become an active field for remote sensing. In 1972, the remote-sensing projects for ERTS and Skylab were listed in two project books, and I counted twenty-eight California projects, sponsored either by the active government agencies there or by the University of California or Stanford University.

But if some of us find California firmly footed in science, others—such as a friend of mine whose boyhood was spent in Appalachia—see it as faddish and unreal. "California can afford the luxury of experimenting with space information," he said. "But if you want real-life, down-to-earth problems, look at West Virginia. They need help and money now. On the ground." To him, remote sensing had all the relevance of a polar bear in the tropics. Just try solving the dilemmas of West Virginia from a far-off satellite, he seemed to be challenging me.

I looked at the project book and was chagrined to note that there was not a single project sponsor located in West Virginia. Continuing to thumb through the book, I looked at the description of various remote-sensing projects. The diversity of problems being attacked by ERTS and Skylab impressed me, as indeed it did each time I studied the book. There was a Massachusetts plan for managing reservoirs, an Oregon study of coastal water pollution, a Tennessee project to detect crop disease, an Argentine study of livestock potential, a Norway project to develop hydroelectric power from snow, an Alaskan plan to monitor proposed pipeline routes. Although I had to tell my challenger there were no West Virginia projects, there were studies listed for forty-four states and thirty-eight foreign countries.

"That's an impressive list, but what is really being accomplished in all these projects?" my challenger continued. "You're making it sound as if NASA were involved in everything from drilling for oil to growing potatoes."

It is, of course, not NASA alone which is carrying out the variety of activities in which satellites offer assistance. A myriad of institutions and agencies are already receiving ERTS pictures from Goddard Space Flight Center and channeling them into useful pursuits. And the number of technicians whose projects will be benefiting from space information within the next decade is, in fact, incalculable. They are an immense team (or "crew," as we will find good reason to call them later) of scientists and practical users of science. Their projects impinge on local problems strung out from Baja California to Maine, from the Arctic to Antarctica—and, indirectly, on a number of such places as West Virginia, as we shall soon see.

Scientists are, for the very first time, intensely inventorying our entire world. They are examining the nutritional sources of the earth with all the new criteria of space age huntsmen, reassessing both cultivable and noncultivable lands and searching out the food potential of the oceans. They are inspecting the world for both conventional and undefined sources of energy for the future. They are seeking ways to use and to conserve our forests and rivers and other resources.

These and many other critical matters will be examined some chapters hence. We will look at a multitude of global dilemmas, and at space age efforts toward their solutions, with the earth-observing satellite emerging as a tool that can provide a unique, over-all view of the world—allowing science to discover exciting new options for a planet with problems.

2. The Ship and Crew

"If I'm elected governor, I'll put a stop to strip mining in West Virginia," promised John Rockefeller IV in his unsuccessful bid to unseat Governor Arch Moore in 1972.

Strip mining—the practice of ripping open long trenches of terrain to scoop out near-surface coal—was a hot issue in West Virginia. More than a thousand miles of streams in the state had been polluted with sulphurous water, and hundreds of thousands of acres of green grass had been ripped open. "Strip mining is like a knife slashing through a painting," Rockefeller had said during his campaign.[1]

However, the more common deep-underground mine, which Rockefeller strongly advocated, was hardly an example of environmental chastity. Contaminated underground water had long been a product of deep mines, and the Appalachian hills were marred with gigantic heaps of coal slag, hauled up from the depths. The piles of slag were often used to dam up mountain streams, but, as demonstrated in the 1972 Buffalo Creek flood, were unreliable when water levels rose; the slag dam in Buffalo Creek had broken, and two hundred persons were swept downstream to their deaths.

The mass funerals which followed the flood were not a new experience for West Virginia. In the past, however, large-scale tragedy had been a result of disasters within the mines themselves; underground coal mining remained the nation's most hazardous industry, and explosions and cave-ins rocked mining communities with disastrous regularity.

Underground mines, in fact, would seem to be a place where

no one with even moderate ambitions for longevity would want to work. Yet thousands of idle men in West Virginia would today jump at the chance to pick up their lunch pails and be lowered down into the ground again. The jobs, however, do not exist, even though national interest in coal as an energy source has been recently rekindled.

Automation in mining has taken its toll of the jobs in Appalachia, and no other industry has come forth to hire the unemployed left in its wake. The poverty pockets on which John Kennedy focused national attention in his 1960 campaign are still there. Since Kennedy's dramatic promise to improve the life of the Appalachian, more than fifty assistance agencies have been cranked into action in West Virginia, with food and antibiotics steadily streaming into impoverished areas. Health of children has improved, true enough, but there has been no economy developed that can sustain that improvement without outside help. After a decade of government investment, the West Virginia hills have continued to offer few jobs, a dearth of farming, a lack of land reclamation, and even occasional gaps in the yields of their "ever-dependable" forests.

The epitome of the problem can be seen in the northeastern part of the state at Dolly Sods, a onetime pastoral and woodland area now reduced to a rugged, wind-blown wasteland. "A transplant from the Arctic circle" is how it appears to American University agronomist Dr. Norman Macleod. In the early 1900s, the timber in Dolly Sods was clear-cut and carried off by railroads. Sparks flew from wood- and coal-burning steam engines, and fires burned everything in sight, including much of the peat soil that covered the area. Left behind was a stark, rocky environment, an unpopulated region which, understandably, seldom attracts visitors.

Most of the problems of West Virginia are like those of Dolly Sods; the causes are deep-rooted and solutions evasive. Land needs to be developed, resources conserved, people employed, mines made safe. New minerals need to be located and new industries created.

"But who can do it? Who can pull all those problems together and solve them?" The question was asked one day by my perennial West Virginia friend in one of our casual round-table discussions.

"Engineers." The answer came abruptly from across the table (not surprisingly from an engineer). "*One job at a time,*" he added.

There was sound basis for what he said, of course. Not only in America, but throughout the history of the world, men with slide rules (or sometimes abacuses) and sweaty brows had provided momentum for man's improvement. In Greece. In Rome. In ancient China. Working before rulebooks were written, engineers were men of action who put their dreams to work while mathematicians and scientists were still cautiously attempting to formulate the laws of Nature. The engineer in our round-table discussion emphatically summed it up. "Engineers built the Seven Wonders of the World while scientists were still *going back for more data.*"

History indeed does confirm the engineer's active leadership, and, furthermore, seems to indicate that it was appropriate for the times as a vast, sparsely settled planet gradually adjusted to the burgeoning levees and ditches and bridges the engineers imposed upon it.

Earth as a Spaceship

Now, however, things have been squeezed closer together by high-speed technology, and the world, in this sense, of course *is* shrinking. Buckminster Fuller has described the earth as a spaceship,[2] orbiting in the solar system in semi-isolation, depending on the sun for her fuel, but otherwise necessarily self-reliant. Spaceship Earth is a craft on which food, like fuel, seems in short supply, and on which all the components of the system are, sometimes perilously, interrelated. Whenever man or Nature acts upon the spaceship in one place, the ship is likely to be affected elsewhere. The geological structures which Nature created in the Appalachian range may determine the cave-ins that will occur in underground mines; the same deep mines, in turn, provide a vital fuel, but also affect the streams and the people living on their banks; and the strip mines with their growing contributions of energy have to be evaluated against the land they consume.

"Anything we want to build now requires an environmental impact study," is the engineer's current lament. After struggling against Nature for centuries, the engineer has recently been informed that he is not to fight her but protect her. He has been instructed not to clutter things up with power plants and freeways, yet to provide ample electricity and speedy transportation.

The engineer is, of course, not equipped to meet these conflicting requests, and Spaceship Earth seems often to be floundering. Her passengers have multiplied to the point that managing her supplies and charting her course is not a task the one-job-at-a-time engineer can handle alone. It requires someone standing behind him with a broader view—and, in a stroke of incredibly good historical fortune, we have the scientist with his new satellite vantage point to do exactly that. Together the engineer and the scientist could function as an operating crew for Spaceship Earth and work toward keeping the craft safe for her passengers. The engineer could propose and design, as he has always done. And the scientist could project the over-all effects into proper perspective, not as the environmental impact on a single plot of ground, but as it would affect Spaceship Earth as a whole—helping the engineer "see around the corner" where additional opportunities and problems might lie.

If the engineer would be disconcerted at having a scientist peering over his shoulder, he would at least have the consolation that the scientist, whom the engineer has accused of historically "going back for more data," *is now suddenly finding it.*

"Why shouldn't we scientists find the data we need?" asked Anthony J. Calio, Director of Science and Applications at the Johnson Space Center (JSC) in Houston, in 1972. "We have access to the information-collecting bonanza of all time—remote sensing. The whole world is literally rotating beneath our microscope—ERTS."

The Spaceship Crew and West Virginia

Appalachia was one small part of the world which found itself revolving under the ERTS microscope in 1972. While no user agency in West Virginia was sponsoring an ERTS project, scientists in neighboring states had requested ERTS pictures of that entire section of the country, with many of their studies relevant to West Virginia.

One such study was being made five hundred miles west of the Appalachian ridge, at the University of Indiana in Bloomington. There, geologist Dr. Charles Wier in late 1972 began examining ERTS data for fractures in the earth's crust in areas around deep-underground coal mines. He was studying six hundred such mines, ranging from some which had been recently operated to others which had been abandoned a century ago. He was noting the relationship of fractures around each mine to the history of cave-ins at the mine, cave-ins being the major cause of death in underground coal mining. If a correlation between fractures and cave-ins could be clearly shown, then satellite pictures would apparently be useful in the future; dangerous fractures in projected deep-mining areas could be located by the pictures. The means would thus be available to guide mining companies (and the Bureau of Mine Safety) in selecting the safest sites for shafts, not only in Indiana but wherever deep mines were projected.

These same geological features being studied by Wier for the sake of safety also lent themselves to various searches for underground wealth. When ERTS pictures first became available, geologists working for mining companies began looking at West Virginia and surrounding areas in search of minerals. Throughout the eastern mountain ranges, from Georgia to New York, they saw lineaments that are now leading to speculation about underground deposits of all categories. In December 1973, one of the Georgia mountain areas which had looked promising in ERTS was reported as a new source of kaolin, the basic material of ceramics. Similar signs of mineral wealth appear in the ERTS pictures of West Virginia.

Oil too had renewed possibilities in Appalachia. The hills of West Virginia had been test-drilled and rejected as a major oil-producing area in years long past. But had this conclusion been completely valid? The ERTS and Skylab pictures now enabled the oil companies to compare the highly productive oil fields of Pennsylvania with land patterns in adjacent states. They could see similarities in the ERTS pictures between the tight folds of Pennsylvania's strata and those of West Virginia. Photo interpreters pored over pictures of such areas, carefully inspecting the terrain for anomalies—unique but subtle features which might suggest the presence of fossil fuels. Whether or not there were any such signs exciting enough to intrigue the geologists was a secret the oil companies would keep to themselves for the present. West Virginia might—or might not—be the place where remote sensing would lead to test-drilling, and test-drilling, in turn, to full-scale exploration. But the pictures were under careful review in a dozen geological offices.

Geologists, however, were not the only scientists studying the ERTS pictures of oil- and coal-laden Appalachia. Ohio State University agronomist Dr. Wayne Pettyjohn found particular interest in the view which ERTS provided of the strip mines scattered over several states. Pettyjohn knew that today, as strip mines were exhausted, mining companies were required to push the soil back into the open strips and replant them with trees. The trick at that point was to place the topsoil back in its original place, on top of the barren subsoil. Otherwise, the trees might just as well never be planted—and, more significant still, the land would be rendered useless, virtually evermore. The more conscientious mining companies therefore were now using elaborate and expensive "Rube Goldberg" varieties of earth-moving machines, which replaced huge chunks of soil the same way they had removed them, top side up. There were, however, other, less obliging companies which were taking short cuts and tumbling the earth back into place upside down. In such cases, the valuable Appalachian topsoil was lost forever under tons of subsoil and overburden, causing the trees, when planted, to die and dramatically reveal the demise of the land.

Now, however, the miners—both the conscientious and the irresponsible—suddenly had a satellite looking over their shoulders as they planted the trees. "By sensing the radiation from the young trees, ERTS can often tell, well in advance, which groves of trees are going to survive and which ones aren't!" Pettyjohn said. The availability of this kind of information (explained in Chapters 8 and 14) is a plum indeed for conservation efforts. "Users in state governments," pointed out Pettyjohn, "can obviously apply such data in a practical way. They can provide mining companies with maps showing if their earth-moving efforts have been satisfactory. They can, in fact, see that bonds are refunded to those companies, and only those, whose trees are going to survive."

In Pettyjohn's project, the observing of tree survival was a means to an end, a gauge to the condition of the soil. In other ERTS projects, however, it is the trees themselves which have been the object of remote-sensing scrutiny, as in the Potomac River Basin study of American University's Dr. Norman Macleod. In June 1972, two months before the ERTS launch, I talked with Macleod not far from West Virginia along a tree-lined lane of Maryland. The maples and oaks which surrounded us, he told me, were in danger of wholesale destruction. Remote sensing from a NASA aircraft the year before had revealed that there was a severe infestation of downy mildew on West Virginia poplar trees. "Downy mildew is serious," said Macleod. "And it's very likely to spread from poplar groves to most other deciduous trees throughout Appalachia," he said. However, he pointed out, ERTS hopefully could be used to create maps of the diseased areas. Then the Forest Service and the logging companies could launch a statewide war against the mildew.

Agronomist Macleod had also inspected Dolly Sods, West Virginia, studying the decline of vegetation there, and developing concepts as to what might be done to reconstitute it. User agencies, he hoped, would employ ERTS data to answer some key questions: Could the area eventually be regrown?

Fifty years before, Dolly Sods had supported livestock grazing. Did it now have the potential to be reconstituted, so as once again to produce milk and meat? More immediately, could Dolly Sods be developed as a wilderness area for tourism, and provide jobs for West Virginians? Would ERTS and Skylab pictures indicate factors conducive for industry to invest and grow and employ? In the years since Macleod asked some of these questions, ERTS has accumulated the data on which a variety of innovative responses might be based.

Elsewhere in West Virginia, ERTS pictures have provided an overview of mountains where ground cover has been destroyed and where heavy rainfall runs unchecked, flooding over hard barren ground. Here, too, users can study the satellite pictures to answer key questions: Can the sparse areas be developed and planted so that the ground can absorb the rainfall and prevent flash floods? Or, are dams really the only means of controlling floods in those areas? If so, today's hydrologists and engineers are looking for technology that will help them determine the required capacities and the best locations for dams. A map of a mountain area showing how much water is stored in a watershed in the form of streams, lakes, ice, and snow would be the first requirement. Hydrologists are learning to determine this kind of information from the ERTS.

A wide variety of relevant studies is thus available for West Virginia. The work of Wier, Pettyjohn, and Macleod holds much potential, and the ERTS data on their studies, which will enable federal and state agencies and private companies to pursue these vital problem-solving matters, will be on public file, available to all.

The studies of theoretical scientists now can be utilized everywhere by practically oriented scientists, engineers, and administrators who are to be remote sensing's link with reality: the *users* of scientific knowledge who indeed are concerned with such matters as "the drilling for oil or the growing of potatoes." Large numbers of such users who are already employing sensing can be found in the Departments of Agriculture and Interior, the oceanographic agencies, half a dozen state governments, and private industry. They range from super-systemized city planners to rugged experts in marine life. What they share is a certain responsibility to get things done, and we like to see them working with facts.

The Scientific Gap and the Steering Committee

Congress, when granting the first funds for remote sensing, had insisted that satellite data be made public, available to anyone who wished to pay the nominal price of a picture*—a policy assiduously adhered to ever since. This decision on the part of the lawmakers may ultimately prove to have been an example of rare congressional insight, no single person or agency being sophisticated enough to comprehend fully all the satellite information. Our spaceship requires the expertise of a variety of specialists if it is to survive; it needs, for example, the energy fuels which miners and oilmen will find lying beneath the green Appalachian forest, but it also needs the conservationist to keep the spaceship operative, rebuilding the forest after the mines and oil wells have run their course. While it would be convenient if scientists were all-wise in all things, they are, we disconsolately find, usually as myopic as the rest of mankind—each of them concerned primarily with the project of his own company or agency. Neither the company chemist nor the environmentalist, if left completely alone, would show much concern for the requirements of the other. One's basic motivation is to pull energy out of the soil without great regard for the aftereffects; the other's is to protect the environment totally irrespective of today's vital energy needs. The crew of Spaceship Earth is often disunited.

If each scientist-crew member is narrow, however, the scope of the satellite is fortunately broad. Potentially it offers all the information needed to run Spaceship Earth. And since no single scientist on our crew can digest all this information and ob-

* Orders can be placed with Earth Resource Observational System (EROS) Data Center, Sioux Falls, South Dakota 57198. Telephone (605) 339-2270.

jectively make decisions, perhaps the answer instead should be to expose as broad a group of scientists as possible to these earth observations. Could a cross section of scientists become, in effect, a "steering committee" for Spaceship Earth, a consortium of talent to examine the earth's resources and problems and to recommend ways to keep the planet alive and well? Scientists of conflicting motivation need to meet and talk and develop concepts that will suit all of the earth's needs. In a limited sense, such a steering committee has begun to develop, with scientists gathering frequently at national remote-sensing symposiums.

Without such broad scientific interchange, man will constantly be attempting projects which may win a particular battle, yet perhaps lose a war. A dam may irrigate a valley, but will it also destroy a bay? Conversely, a policy controlling river vegetation may protect a wildlife refuge, but will it waste unreasonable amounts of water and energy? A supersonic transport may move at exciting speed, but will it or will it not excessively pollute the stratosphere?

To answer these kinds of questions rationally rather than emotionally, the scientists will continue turning to remote sensing "for more data" on interrelated projects. When an engineer proposes an "Eighth Wonder" for the world, our steering committee may agree that it should be built, to keep Spaceship Earth functioning smoothly. Or that it should be pre-empted because it is judged harmful to our craft, likely to waste more food or energy than it can develop.

Contaminants in the air led us to the wondrous smog devices on 1973 cars, but, unhappily, the devices in turn brought us to a staggering consumption of gasoline. A similar concern over air contaminants has caused us to disdain grimy factories and, therefore, to develop more interest in nuclear power—only then to find a strong public reaction over alleged thermal pollution from nuclear reactors. Such circles of confusion are not confined to these examples of air and fuels, but encompass most of the other problem areas of this beleaguered earth. The complexity of the earth's problems demands that they all be measured and documented, and often interrelated in a variety of experiments by remote-sensing satellites.

We started this chapter talking about just one of the fifty United States, and we finish by acknowledging what we knew all along—that the world and its problems are substantially larger than West Virginia. Part of the crew on Spaceship Earth can and should be using remote sensing to study the problems of each locale, in every activity which they deem worthy; but all these local efforts then must be coordinated into plans for the earth's survival. If our critical problems of supplying and maintaining Spaceship Earth are to be solved, our first job is to see that satellites provide our leaders with an over-all picture, showing where both our difficulties and our resources lie.

3. The Tools and Where They Came From

Before plunging into the exploration of the world from space, it might be wise to spend a chapter examining the tools of the sensing trade. This makes it necessary to acknowledge forthwith that remote sensing is really much more than simply "taking pictures." Sensing began long before the camera was invented, and it has recently gone far beyond the scope of the camera. Sensing actually refers to the ways that man detects and identifies all the objects in the world about him; *remote* sensing, then, has to be the identifying of those objects which are some distance removed from him.

The sensing technology as we know it today was fused into existence during the Apollo era, without anyone being aware of just when it happened. By the mid-1960s, it had all come together, and science soon began to recognize its significance. Like most technologies which are called new, it incorporates some concepts as old as the sun. In fact, remote sensing utilizes the sun itself as its direct source of power. The rest of the system, however, is considerably newer, having existed only since the advent of the human eyeball.

The sun and the eyeball. These are the components of remote sensing. And we can probably appreciate the whole concept more fully by going back to the days when these two units were, as aerospace people are prone to say, interfaced into a system.

It started, some millennia ago, with a man we may just as well call Adam, who came to the realization that his body was equipped with an eyeball. And an eardrum. The eardrum, in fact, did a lot of the sensing for Adam in the beginning, listening to those provocative whispers from Eve. But suddenly it was the eyeball that came into play, focusing on a bright red apple. And Adam's eye recognized the apple as something special.

Here's why.

As Adam looked at the apple, millions of light rays from the sun were bombarding it. They were, in fact, streaming down on everything in sight, including Adam, and have tirelessly continued this downpour ever since.

It takes eight minutes for a photon to leave the sun, as part of a whole train of photons making up a light ray, and to arrive where you, the reader, are sitting. This very page is being deluged with photons (from sun or artificial light), with many of the rays bouncing off in all directions, some of them zinging into your eye. The black ink on this page absorbs most of the rays that strike the printed characters themselves, releasing relatively few rays to bounce toward you, whereas the white space reflects most of the light that hits it. This means that these bouncing rays are forming an image on the retina, in the back of your eyeball, that is a reproduction of this page. With extraordinary efficiency, you have sensed black from white.

As soon as these black words register on the retina of your eye, they are flashed to your brain—an impressive storehouse of knowledge, including a catalogue of words which it and you have encountered before. As word after word is sent in from your retina, your brain works at lightninglike speed to compare the words you are reading with the "signatures" of familiar words already on file in your brain. Each word on this page is thus identified, and the meaning recognized.

Adam's experience with the apple was much the same as yours with this printed text, except that he was viewing a scene in full color. Adam's eye did what television has been doing more recently—separating the apple into three images, each a primary color. One image of the apple turned out a light red, another pale blue, the third a washed-out green.

Once on the retina of Adam's eye, the three color separations were reblended into the deep natural near-red, duplicating the actual color of the apple perfectly. The full-color image then sped to Adam's brain. While it was the first time Adam had ever looked at the apple, he had become acquainted with a gardenful of other fruit, and his brain had an impressive catalogue of fruit signatures to try to match up with the apple's image. The apple was close enough to, let's say, the pear signature so that Adam's brain had no trouble cataloguing it as a similar piece of fruit, yet different enough to require that a special new apple signature be created and left on file.

Adam's other senses, like his eyesight, had swift lines of communication with his brain, so that as he touched, smelled, and tasted the apple, his brain went through the same ritual of shuffling through its inventory of signatures, finding similar ones, and finally having to chalk up a new one.

Thus works the system of natural sensors. It worked so well for Adam and his primeval descendants that no one found any need to improve on it for a while. Three of man's senses, in particular, were telling him about everything happening for a reasonable distance around him, and this apparently kept him satisfied; he could generally see, or hear, or sniff his prey, his enemies, and even the objects of his affection, and his brain managed to keep some semblance of a record as to how things looked, sounded, and smelled.

Early Man-Made Devices

As history inched along, however, man apparently developed a medieval case of cabin fever, and decided to find out what was going on beyond the range of his senses. He achieved this variously through the use of telescopes, thermometers, weather vanes, audiphones, barometers, and seismographs. The awareness they provided would have burdened his brain with the chore of storing the information, but the brain, too, was gradually getting outside help from a parade of innovations—stone tablets, scrolls, paintings, graph paper, and eventually the printing press. By the nineteenth century, new devices for both obtaining and recording information were coming hot and heavy; telephones, telegraphs, phonographs, and the modern camera were collecting, transporting, and storing knowledge in ways which would eventually affect every aspect of life.

As if this inventive explosion weren't enough of a revolution, man began to take to the air in the wonderful age of ballooning. The eyeball was being elevated and exposed to a panorama a hundred times greater than anything it had experienced before; old maps and concepts that had seemed axiomatic suddenly had to be torn up and redrawn with new perspective.

The Camera in the Air

Enter the military. In the late 1850s, a French army officer took an aerial photograph by remotely controlling a camera in a kite, and flight and reconnaissance were at that point irrevocably married. Just a few years later in the United States, the Union Army began using balloons for photographic scouting, planning the battle of Richmond from a single detailed photograph that showed not only the southern troops, but also the rivers, railroads, swamps, and woods, with an accuracy never before known in mapping.

There was no doubt about it: the camera people were on the move, and by the twentieth century they were ready and waiting for the flying machine and the opportunity to be transported by it. At the beginning of World War I, the airplane was looked on as an opportune means of scouting, of extending the range of the eyeball and camera. Neither the Germans nor the Allies in 1914 recognized the airplane as a carrier of weapons. German and English aviators observed and photographed troop movements and gun emplacements from unarmed planes, and sportingly waved at each other as they met flying to and from their respective reconnaissance.*

But as its applications increased, the airplane evolved rapidly,

* This gentlemanly use of the skies continued until pilots began carrying rocks and pistols, and, alas, finally bombs and machine guns.

both throughout the war and in the 1920s—so rapidly that now the camera had to strain to keep pace. Air speeds began to advance faster than shutter speeds, and the photographic industry on both sides of the Atlantic struggled to avoid blurred aerial pictures. They were soon successful enough so that sharp photographs not only were available in quantity but were being pasted together into great photographic mosaics. These startling aerial overviews began to seem important to a number of folk. "The nation with the best aerial pictures will win the next war," Germany's General Werner von Fritsch predicted.[1]

The early 1940s proved the general correct. George Eastman's country and the victorious Allies photographed millions of square miles with sensitive films and with cameras that ranged from hand-held types to sophisticated clusters of lenses mounted on aircraft.

Rays the Eyeball Doesn't See

General von Fritsch and others recognized that the camera had perhaps its greatest potential in "seeing" things which no eyeball could detect. For despite all our comments about the unique abilities of your eyes, they actually can distinguish a scant 3 per cent of the rays which stream in through the pupils. These relatively few rays are visible because only they vibrate in wave lengths the human eyeball finds compatible. Ever since Adam, we have enjoyed a rainbow of colors, caused by the varying wave lengths of the visible rays. Our eyes can see the violet rays at one side of the rainbow, but cannot quite manage the ultraviolet that are just beyond; at the other side, we can see the red, but not quite the near-infrared.

The ultraviolet (UV) rays, best known for their tendency to broil overexposed human bodies, dart busily around us throughout our daylight hours; but they come in wave lengths too short for our eyes to see. We are equally oblivious to infrared (IR) rays, whose wave lengths are too long for us. IR rays are the conveyors of the heat from the sun—or, for that matter, from any source, including the warmth emitted by your body, by the furnace in your home, and by the red heat lamps used for aching backs the world around. Some of the UV and *near*-IR rays (see Figure 1) bounce off objects just as the visible rays do,

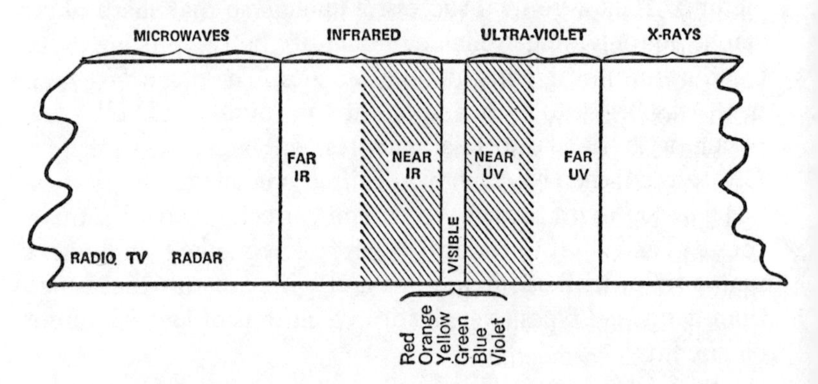

Figure 1: Partial sketch of electromagnetic spectrum. Man's limited range of visibility (center band) can be extended broadly by sensors which operate in nonvisible wave bands and "see" unique characteristics of the earth imperceptible to the eye.

zinging into our eyes, but since we cannot see these rays we fail to observe certain qualities they reflect from the various objects. We thus constantly fail to see a wealth of information which is bouncing toward us.

(A certain portion of the sun's rays, however, instead of being bounced off the objects they strike is absorbed by them. Their energy is converted to heat by the objects and steadily re-emitted as far-IR rays, *day and night*—a most significant trait. If you could miraculously attune your eye to these far-IR rays, you could, even on a night that was black as pitch, see a large share of the world around you.)

While our eyes will never directly see UV, IR, or any other rays outside the visible range, the camera has no such built-in limitations, and from the beginning it has seen some things which we could not. While the early-nineteenth-century film failed to respond to reds, oranges, or yellows we ourselves could see, it nevertheless detected the UV rays which we could *not* see. Persons subjected to some of the early-nineteenth-century portrait photography were understandably dismayed with the results. Instead of picking up the rosy tones in their cheeks, the film unveiled splotchy variations in skin pigment, and turned the UV luminescence from teeth into fiendish grins. The camera wasn't lying, but it was uncovering a rich new ultraviolet dimension to life. (If it was frustrating to portrait photographers, it did hold promise to dermatologists.)

The real pot of gold, however, lay at the opposite end of the rainbow—the red end—although it was well into the twentieth century before IR rays were to be utilized to any degree. In 1931 a group of men from the U. S. National Research Council were gathered in a room in Rochester, New York, at the Eastman Kodak Company, discussing the state of photography. At one point in their meeting, Kodak's Dr. C. E. K. Mees nonchalantly asked that the group allow him a few minutes' break in the proceedings for a minor experiment. The lights then went out, and the tightly curtained room was plunged into total darkness.

The shutter of a camera silently opened, and closed one second later.

The lights came back on. Most of the group had guessed that an exposure was being made, but none of them had been aware that they were being bathed in near-IR light which bounced off their faces and bodies to record their images on infrared-sensitive film. "You have just been part of the first group picture taken in total darkness," Mees informed them, proving it an hour later by handing a print to each of them. (Other experiments of the same era involved red-hot flatirons, which emitted their own heat, and, without spotlights, were recorded on far-IR film.)

"Dr. Mees demonstrated the camera's ability to see IR radiation in darkness," remembers Kodak's Dr. Walter Clark, who stood in that dark room in Rochester more than forty years ago. "But in 1935 a more practical demonstration of IR occurred in a tremendous achievement." The National Geographic–Army Air Corps balloon, Explorer II, ascended to 72,395 feet, from which point its two occupants looked toward a hazy northern horizon 330 miles away. One of the men, an army photographer, poked a cannonlike lens over the side of the balloon gondola and began taking pictures with different films and filters.

Among the pictures taken, the IR film proved to be the best, and its clear reproduction of the earth's curved horizon demonstrated several things. First, for any who still had doubts, the world was round. Less predictably, it showed that IR rays could cut through thin haze and record a picture at 330 miles, and could reveal some aspects of vegetation that went unnoticed by regular film. (The military, especially in Germany, was to improve on this ability and in World War II used color-IR film to distinguish true vegetation from camouflaged buildings.)

Once photography had shattered the red barrier, other sensing devices were not long in following the example and expanding upon it. Just beyond the IR band in the spectrum (see Figure 1) are the microwaves, a band of waves only sporadically emitted by the sun, but utilized regularly in "man-made rays" generated by radar, radio, and television. By using these electronic devices, man has become more "active" in his sensing; whereas the camera and the eyeball usually receive only the rays showered around them by the sun, the radar makes its own signal, beaming it toward objects its masters wish to detect. The signals are then reflected back to the radar and identified by it, just as the sun's rays are recognized by the eyeball.

Throughout World War II the performance and use of radar advanced steadily. While radar lacked the quality of a camera, it could see through the night every bit as well as near-IR film. And the radar microwaves could penetrate a sky full of rain or pea-soup fog that would have stopped even near-IR film like a brick wall.

The Cold War and Nonphotographic Sensors

Throughout the early 1950s, a series of U.S. naval and air force aircraft were shot down near the periphery of Red China and other Communist countries. The fact that they were carrying sensing equipment remained for then a superbly kept secret. But in 1960, when an outraged Nikita Khrushchev waved a shoe in the United Nations in response to CIA pilot Gary Powers' U-2 flight, cold war "snooping" became an undeniable fact of life. The *Pueblo* incident a few years later created another dramatic example.

Even today we don't know all the equipment the U-2 and the *Pueblo* carried. But we do know that the cold war was a time when cameras and films were improved. And we now know that the same years also advanced us on our way to a new generation of sensors which were "nonphotographic," *yet capable of supplying cameralike pictures.*

These exciting nonphotographic sensors could absorb radiation from a subject and convert it into an immediate "image" that looked like a photograph but could be relayed from sensors in airplanes to ground stations, much like television signals. These sensors were able to operate in more of the spectrum than anything that had come before, ranging from UV to visible, to IR, to microwaves. Scientists found that sometimes by trying out several wave lengths at once they could unearth a variety of totally new information. A man with his eyeball trained on a hillside, with the most penetrating gaze he could muster, might see only a pastoral landscape—while an IR sensor trained on the same area could perhaps detect soil moisture patterns warning us of a potential landslide! And a UV sensor viewing the hill might see hints of a feldspar of aluminum. Each of the three images obviously has special value; ergo, the more wave lengths the merrier.

Taking this shotgun approach, scientists began developing a variety of multiband sensors, which were carried by airplanes over the next several years in an effort to learn just what might be seen. The sensing was carried out in a dozen or more wave bands, in attempts to inventory such varied matters as the growth of farm crops, the health of trees in forests, and the characteristics of waterways and marine life. Generally, the sensors were quite effective; new information about the world was being accumulated faster than man was able to learn to interpret and utilize it.

An Outside View of the Earth

But even as scientists began to progress in interpreting their new source of knowledge, another new combination of technologies was developing that would completely revolutionize the young world of sensing. "There were three new surges of technology breaking in the mid-1960s that eventually affected remote sensing," according to Anthony J. Calio, the Johnson Space Center science director. (It was, however, to be several years before the three would develop and be merged into a single technology.)

The first of these three surges was the elevating of a sensor well above the earth's lower atmosphere for a true overview of the world. A significant example of this had occurred nearly ten years before Sputnik I, on July 26, 1948. On that date, a V-2 rocket was launched from White Sands, New Mexico, to an altitude of sixty miles (five times the height of the 1935 army balloon), with a camera automatically snapping pictures from its nose cone. The half dozen pictures, when patched together, provided a mosaic hundreds of miles wide with reasonably good detail; previously such a mosaic would have required several hundred aircraft photographs. In retrospect, at least, the event seems to have been of monumental import.

Apparently, however, the few scientists who looked at the mosaic were more interested in rocketry than photography; the picture was dropped unnoticed into government files, allowing the scientific community to remain generally unaware of remote sensing's potential for several years to come. Indeed, throughout the 1950s, when satellites were first becoming a believable concept, meteorologists were the only scientists interested in space pictures—and their early interest is understandable. Weather seemed easy to photograph. While space pictures would be (everyone presumed) too blurred to provide detail of the earth itself, no one doubted they would be accurate enough to provide outlines of weather fronts hundreds of miles wide.

In 1955, a U. S. Navy Viking rocket which was shot up from White Sands, New Mexico, brought back another series of pictures, each covering thousands of square miles, and the meteorologists' interest in cloud study was reinforced by a glimpse of Los Angeles smog trailing off into the Mojave Desert. (The San Andreas fault, of earthquake fame, appeared distinctly in the same picture, but, strangely enough, caused no great stir in the geological community.)

Unlike the geologists, the meteorologists reacted, and in April 1960 the first weather satellite, Tiros I, was launched. Nine days later Tiros dramatically revealed her merits to the world by telemetering a fuzzy but unmistakable picture of a typhoon forming near New Zealand. The ability of the camera to provide advance weather warning was suddenly apparent, and the 1960s became a period of weather-satellite experimentation. The public continued to look on daily weather forecasts as more hilarity than science, but meanwhile satellites were becoming truly impressive in their ability to detect tropical storms. Satellite warnings in 1967, and again in 1969, were principal factors leading to the timely evacuations of tens of thousands of Gulf coast residents just hours ahead of devastating hurricanes Beulah and Camille.

Geology—The First Breakthrough

Meanwhile, NASA was shifting into high gear in its drive to the moon, President Kennedy having initiated both the verbal challenge and the federal budget. In September 1961 the fourth Mercury capsule was put into space, unmanned but carrying an automatically controlled reflex camera with 70-millimeter film.

The photographs which came back on Mercury 4 were amazingly sharp. They included some striking views of French Morocco and were greeted with a chorus of exclamations from NASA and the public. "Gee whiz" photographs was the name NASA people gave to these pictures which, like the snapshots of any tourist, served no apparent purpose except to intrigue the viewer. The response by the scientific community was, once again, casual, with most of the reaction to the pictures coming from enthusiastic laymen who saw them in *National Geographic* magazine.

Among the readers of the *Geographic*, however, was a young geologist, Paul Merifield, then working on his doctoral thesis at the University of Colorado. "I was impressed with the quality of the pictures, and equally impressed with their content," Merifield told me years later. "The folds in the Moroccan mountains showed how strata had been disrupted and re-formed. They were pictures that gave you a million years of history all at once." He paused, then added, "They also contained intriguing hints of oil deposits."

Through Merifield's work, science began to recognize that satellites had made a stupendous breakthrough in the study of the earth. Merifield, in scrutinizing the Mercury 4 pictures alongside a French geological map of the area, found a strong similarity between the map—which had been methodically developed over a period of decades—and the remarkably sharp two-inch squares of film! He was sure that the pictures would be well defined enough to use in mapping, and more accurate than any conventional map could be. "The French," said Merifield, "had arrived at a number of accurate conclusions about the area as a result of their thousands of miles of tedious hiking over the years. But now, much of the same information had suddenly been obtained, and even improved upon, by satellite—with the single click of a shutter."

The Bottom of the Priority Scale

Merifield used the Mercury photos and the earlier White Sands pictures to develop his doctoral thesis, and in the process he contacted another young geologist, Dr. Paul Lowman, who had then recently been hired by NASA to coordinate moon photography. He pointed out the value of the pictures to Lowman, who was quick to concur. "Merifield convinced me that NASA should be taking pictures of the earth, as well as the moon," Lowman recalls. "He planted the seed of remote sensing years earlier than it might otherwise have been sown."

By then it was 1962 and John Glenn had orbited in space, carrying along a 35-millimeter camera and returning with more "gee whiz" earth pictures. Lowman, although then in a minor role at NASA, personally asked subsequent Mercury astronauts to do likewise. Scott Carpenter complied, trying valiantly on his three-orbit Mercury 7 (already a nightmarishly crowded mission), but brought back a poor assortment of pictures. Next was Wally Schirra, with the six-orbit Mercury 8. His mission was almost as crowded as Carpenter's, and Schirra, never a deeply inspired photographer, managed to overexpose everything he shot.

Then Gordon Cooper was launched on his 22-orbit Mercury 9, a marathonlike trip by 1963 standards. Cooper found considerably more time to experiment with photography on his mission, and about half of his results were good pictures. Among them was an outstanding shot of Tibet. It was so clear a photograph that Lowman used it to draw the first known map of that part of the Asian range. Geographers took note.

Lowman had by then succeeded in having all the astronauts in the program supplied with 70-millimeter cameras and taught how to use them. Although he found this cooperation encouraging, he was operating in a period when every moment of astronaut time and every ounce of weight in the space capsules were calculated. Neither Lowman nor anyone else had any illusions about how photography ranked on a NASA priority scale. "If anything were going to be thrown off a mission, it would be a roll of film," said Richard Underwood, another NASA photo specialist.

The next extensive manned flight was Gemini 4, scheduled to last four days, and Lowman and Underwood began months ahead of time to promote a photo plan for the mission. As a result of Merifield's work, geologists and geographers in large numbers had joined the pursuit of space photography. Lowman briefed Gemini 4 astronauts Edward White and James McDivitt on several areas in which geologists had asked for photographs, and he also encouraged the astronauts to shoot at additional targets of opportunity anywhere along their course. Because of its cloud-free weather, North Africa was to be a major target area under the orbit path, as it had been for Mercury 4.

At this point, NASA was developing at least a temporary enthusiasm for photography, and with good cause. Gemini 4 was to include a photo mission with considerably more popular appeal than the shooting of earth pictures: Ed White was to take his spectacular space walk. The Soviets had just announced the completion of a space walk of their own, and had released a picture showing the head and shoulders of a cosmonaut supposedly outside his capsule in space, but there was some doubt as to the picture's authenticity. NASA leaders were determined that there would be no question about the U.S. walk; while White was taking his historic stroll, McDivitt would be assigned to shoot pictures of him and thereby document the walk beyond a shadow of a doubt.

The operation went as planned. White, the space walker, and McDivitt, the photographer, both completed their assignments effectively; a set of spectacular color pictures captured the jubilant White again and again as he floated weightlessly through the sky.

Shortly after the return of the Gemini 4 crew, recalls Underwood, the pictures were processed and viewed by photo people and by top NASA brass at Johnson Space Center. Dr. James Webb, then director of NASA, and Dr. Wernher von Braun were among the assembled viewers. They were clustered around one end of a long table, "ohing" and "ahing" over the dramatic pictures of White. Dr. John Brinkman noticed Underwood at the other end of the table, studying the strips of film that contained the earth pictures White had taken during calmer moments of the mission. "Hey, Dick," Brinkman called ebulliently to Underwood, "come take a look at the action."

Underwood's reply, grunted from behind his stack of earth pictures, was classic. "The real action is down here."

The "Synoptic View"

Pictures of the earth were not destined to divert any of the world's attention from the dramatic space walk, at least not at the time it occurred. But, as Underwood implied, the earth pictures were possibly the more truly significant "action," at least insofar as they would affect the earth's inhabitants.

The earth shots were many and varied. Ed White had snapped one of the pictures at the beginning of a film cartridge, using up that first portion of film which photographers seldom expect to turn out. He "wasted" that shot on a dull strip of desert in north-central Africa. The result of this happenstance shot was a detailed picture of the sand dunes, revealing heavy grooves running through the sand for hundreds of miles in a pattern which startled geologists and stimulated new theories about the formation of the dunes.

Another set of pictures extended along the orbit path from Baja California to Texas. These two dozen satellite pictures, of course, embraced an area which previously could only have been covered by thousands of aerial photos. This was impressive, but much more significant was the fact that the photos revealed a network of faults which until then had been completely unrecognized by geologists. (Faults tend to run underground, unnoticed, for miles at a time, and then surface again.) Features which had appeared as short, insignificant gullies when viewed one at a time, either from the ground or from airplanes, suddenly became meaningful when strung together in one great satellite picture. The earth was apparently covered with a labyrinth of giant cracks, relatively few of which had previously been known to man. "At one time I camped right on that fracture. In fact, I probably slept on it, and never realized it," was one startled geologist's response to the amazing Gemini pictures. The suddenly recognizable fractures in the earth were excitedly studied by geologists involved in the search for underground water, oil, minerals, and, especially, earthquake-prone faults.

The "synoptic view," as Lowman liked to call the new broad overview from space, was proving itself in geology and, to almost as great an extent, in geography. The voices of Lowman and others who had considered themselves crying in the wilderness now found they were part of a growing chorus of scientists calling for more pictures of the earth. And their surge of interest opportunely came at a time when the Gemini period was to be succeeded by the bigger and better Apollo era.

"I am the eye through which the earth beholds itself," the Apollo of Shelley's epic poem had grandly stated as he gazed earthward from his chariot in the sky of Greek fantasy. But it was not until the Apollo of our later age, in a situation quite apart from fantasy, that the eye of Apollo would be offered to an earth capable of "beholding itself." The Apollo satellite program of the late 1960s was, of course, moon-oriented, and only two of the manned spaceships—Apollo 7 and 9—were to be earth-orbiting. However, the Gemini pictures having at last awakened the world of science, earth photography was all at once recognized as important. Orders at NASA were passed down to take abundant pictures, and the Apollo astronauts responded. Wally Schirra, the errant photographer of Mercury 8, was the Apollo 7 commander, but this time he came back to earth with hundreds of effective photographs.

The Apollo 7 results were interesting enough so that a more sophisticated system was developed for Apollo 9. Instead of hand-held cameras, the capsule was mounted with four 70-millimeter Hasselblads, offering simultaneous pictures in different color bands of the visible spectrum. The green band was used to enhance features in the oceans, the red was effective for detecting man-made structures, and the near-IR brought out agricultural crops and forests. Discoveries concerning all these subjects were made in revealing pictures, some of which would remain the object of study for years. The earth was beholding itself, again and again, in thousands of photographs scattered over the many faces of the globe. And enough was being learned in these sporadic pictures for science to be hungry for something more comprehensive.

Fusion and the Modern Era

Johnson Space Center's Anthony J. Calio had mentioned that three technologies had revolutionized sensing in the 1960s. One of these had been satellites. The other two, like satellites, had been developed largely for the manned space program. One of these was the electronics explosion; electronic optical and mechanical instrumentation was being developed—with transistors and solid-state components. Thus, the nonphotographic types of sensors which had been created for airplanes during the cold war were suddenly available in transistorized proportions. A sensing device that had been the size of a mattress was now reduced to the dimensions of a pillow. Elaborate multispectral scanners could now be tucked neatly into satellites.

The third surge of technology to which Calio referred was the refinement of computers and the veritable explosion of their capacity. Adam's brain had long before provided the technology to be imitated in the modern computer; as sensing data flowed into the electronic brain, it was matched against the signatures already memorized in its data bank. Fortunately for remote sensing, the moon race of the 1960s had demanded a phenomenal expansion in computer activity. The Gemini-generation computers had given way to the Apollo generation, with a tenfold increase in capacity. Developed for the moon race, the new equipment was now conveniently available for earth observations.

The Earth Observations program now had the triumvirate it needed—satellites to scour the earth; sensors small enough to be carried in space, yet capable of rapid data collection; and electronic brains to store the knowledge. The fusion of these three factors, which had occurred sometime late in the decade, became formally recognized in 1969 when Congress granted funds for ERTS-1.

This was the green light for which the scientific community had been waiting. Hundreds of projects would be planned and executed in tremendously comprehensive fashion. "Remote sensing" and "earth observations" would be facts of both life and vocabulary. The cameras of the Apollo era were in large measure being replaced by versatile sensors, which, when joined with Apollo computers, were capable of compiling data on tape and of storing and displaying it in innovative systems. Where an agriculturist might previously have studied an Apollo space photograph with a magnifying glass, attempting to distinguish a field of soybeans from a grove of pine trees, he could now look forward to an ERTS digital printout in which each dot could unmistakably indicate a soybean field, each dash a grove of pine trees. Or, even better than a computer printout, the agriculturist might feed the tape into an electronic viewer, and then order it to project the soybean fields on the screen in bright green, and the pine groves in shocking pink. "We won't be *looking* at pictures any more," one technician explained it. "We'll be analyzing them electronically and precisely."

Indefatigable ERTS Covers Earth with Three Sensors

"Comprehensive" is another word the technician might have used to describe the ERTS picture. Instead of the scattered photographs available with Apollo, ERTS offers continuous sensing of the globe. In its first year, it compiled literally millions of times the information collected by the combined Apollos. Except for a small blind spot near each pole, every point on the earth's surface is vulnerable to ERTS's surveillance each eighteen days. The eighteen-day interval is the time required for ERTS to cover the world fully with its continuous 100-mile scanning path.

ERTS was designed for polar orbit, traveling from the Arctic to Antarctica on the daylight side of the earth, then northward

Figure 2: By obtaining consecutive pictures every twenty-five seconds, ERTS covers the world every eighteen days. Drawing courtesy Hughes Aircraft Company.

again on the dark side, requiring 103 minutes for the entire orbit. As ERTS arcs northward and southward, the earth rotates east to west within the satellite's orbits. ERTS's track, indicated in Figures 2 and 3, spirals entirely around the globe each day. The following day it will fly a parallel set of paths—each 100 miles to the east of today's tracks. It completes a total of seventeen such parallel sets of paths and, on the eighteenth day, commences the cycle again, having rearrived over the course followed on the first day.

The equipment to perform this demanding task is no larger than a Volkswagen. ERTS weighs only 1,965 pounds, with an over-all body length of ten feet. It was built by General Electric Company. Two solar panels, which receive rays from the sun for conversion into the required electrical power, are positioned on the sides much like large wings, giving the spacecraft the appearance of a giant butterfly. Its eyes are three sensor units at the earthward end of the body.

The first sensor, the Return Beam Vidicon (RBV), is a supersensitive television camera. At any given moment, its field of view is a 115-mile square on the earth's surface. Like an ordinary color television system, it views the same scene in three color bands (one of which, unlike TV, being a near-IR band). Twenty-five seconds later, when ERTS has moved 100 miles farther south, the RBV views another 115-mile square (fifteen miles allowed for overlap).

The second ERTS sensor is a Multi-Spectral Scanner (MSS), literally a rocking mirror which scans horizontally (east-west across the ERTS track). Whereas the RBV absorbs an entire 115-mile-square instantly, the MSS rapidly scans one line at a time, continuously sweeping back and forth across the ERTS track below, covering a strip 115 miles wide, to coincide with the RBV. The scanning is simultaneously conducted in four color bands. The bands are very similar to the four carried in the Apollo camera cluster—one green, one red, and two near-IR bands. (As it turns out, the MSS made by Hughes Aircraft Company, has done the lion's share of the work for ERTS. Because of an electrical problem which occurred shortly after the launch, engineers found it advisable to hold the RBV in reserve for most of the ERTS mission.)

When sensing the United States, both the RBV and the MSS can transmit their data "live" to one of the three ground stations (Maryland, Alaska, or California), almost simultaneously with the sensing. The sensors are also hooked up to video tape recorders, capable of storing picture information ("imagery") for delayed readout. Data from sensing conducted well outside North America (where ERTS cannot make transmissions to the three ground stations) can be stored on tape and then transmitted as soon as ERTS next approaches North America.

All of the sensing conducted by the RBV and the MSS can be utilized in the various forms discussed earlier—pictorial, computer printout, or electronic display.

The third ERTS sensor is the Data Collection System (DCS), which in reality performs more of a communications function than a remote-sensing function. The DCS is being used in a handful of ERTS projects related to various natural disasters, where vital information is collected from remote areas on the ground. Critical measurements of flood levels, for example, can be collected on the ground by a large number of unmanned gauges and relayed, via ERTS, to a flood-watch center. Whereas the RBV and MSS can sense a particular point of land only once every eighteen days, the DCS can monitor a gauge several times a day. Having only to pass close enough for line-of-sight communication, the DCS can thereby receive a message, sent from any point in the world, and place it in the hands of users less than twelve hours from the time measurements are made on the ground. The potential saving of life and property is enormous.

While ERTS's position relative to the earth is constantly changing, it relentlessly follows the sun. Whenever it is on the daylight side of the globe, its westward-slanted path (see Figure 3) allows it to remain ahead of the sun. On the daylight part of the orbit, it is always in virtually the same position relative to the sun (giving it the designation of a modified "sun synchronous" satellite). As ERTS passes over each successive earthly area, the angle of the sun, and therefore the local time of day,

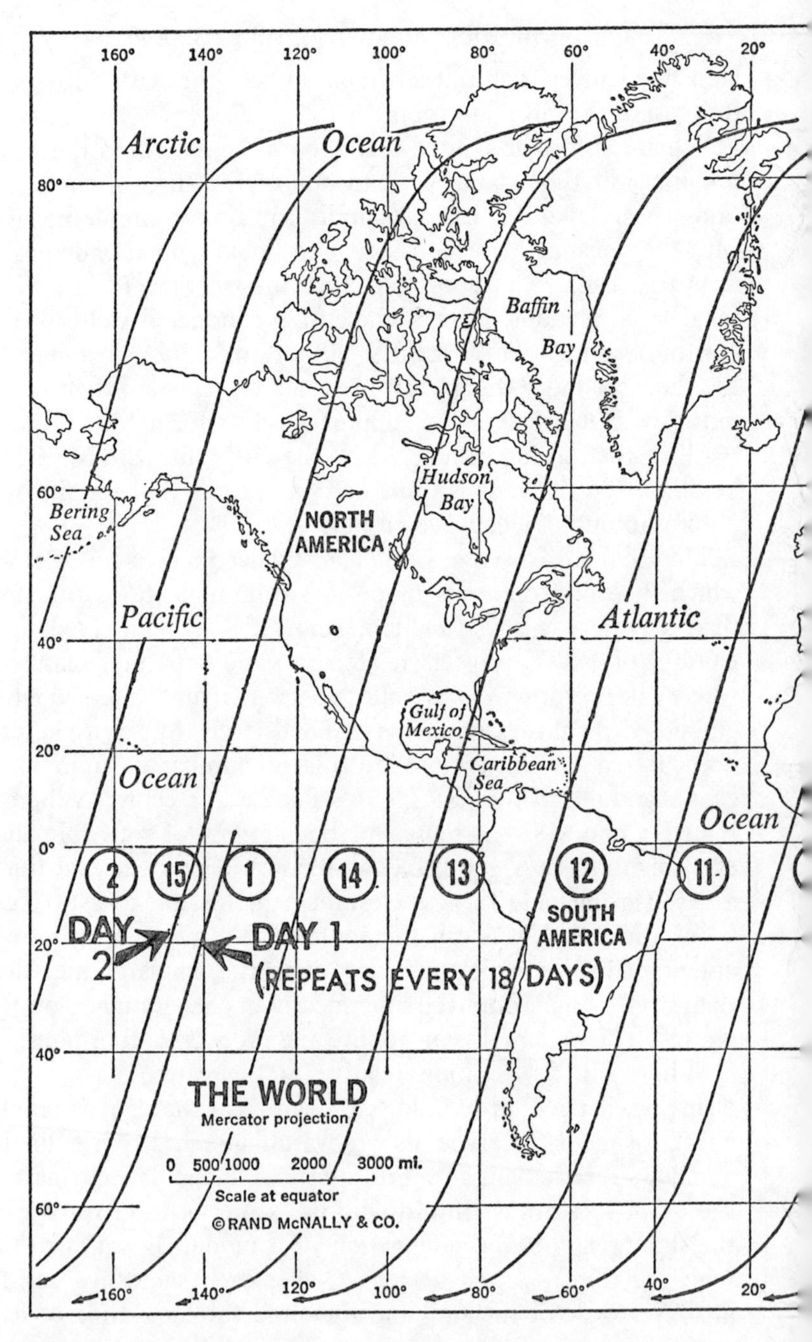

Figure 3: ERTS tracks southward (and slightly westward) on the sunny side of the earth, as the world rotates from west to east inside the satel-

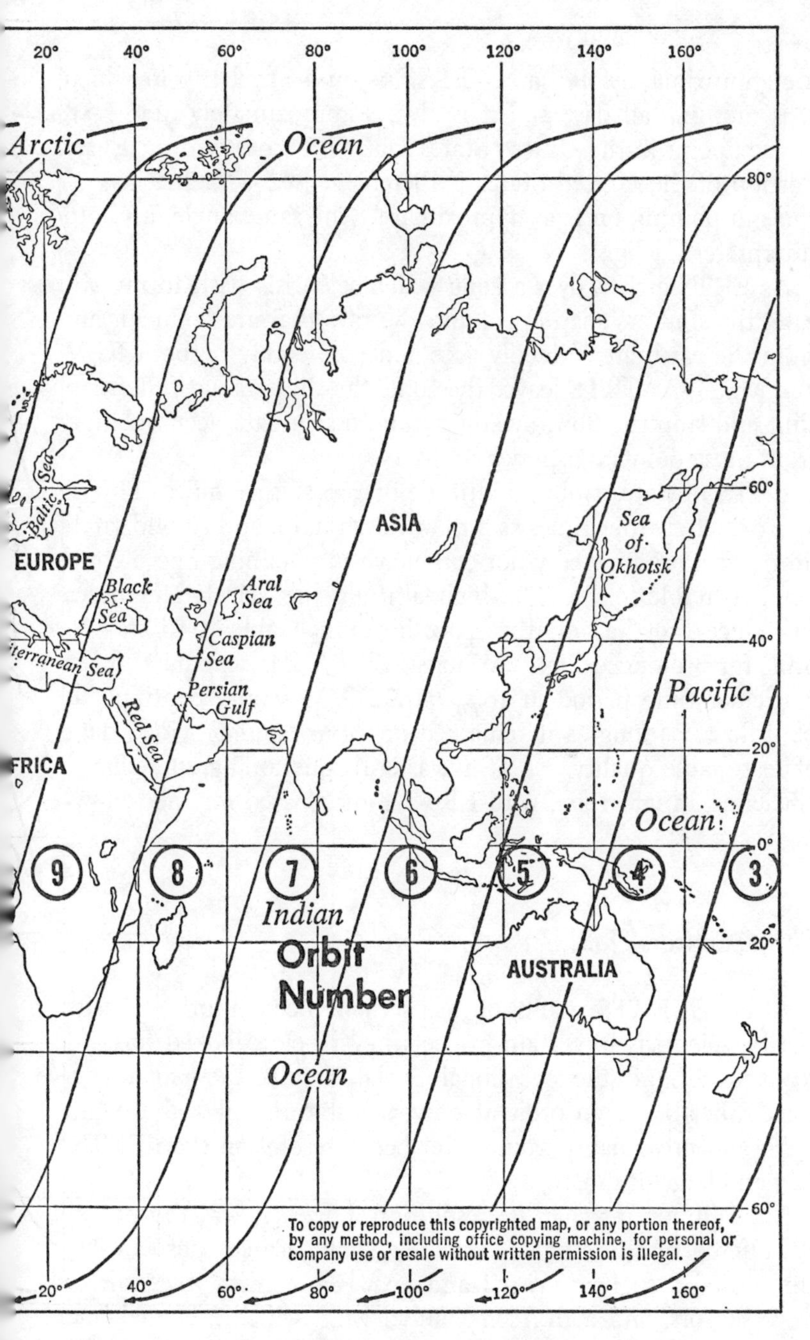

orbit. Drawing shows a typical day's track of southbound passes
Tracks for the following day will be offset 100 miles to the west.

are approximately the same—ERTS is on a novel trip in which it is morning all day; and it is always approximately 9:30 A.M. at each point in the United States that ERTS passes over. Every portion of the United States is thereby sensed with the rays of the sun pouring on it at approximately the same angle, a photo interpreter's delight.

As ERTS arrives over a zone which scientists wish to investigate, the shutters snap open, and the rays that are bouncing up from the earth are suddenly accepted (preferably by both RBV and MSS). As ERTS leaves the zone, the shutters dutifully snap shut and stop the flow, avoiding unnecessary data accumulation from those points not under study.

ERTS operates from an altitude of 570 statute miles. This is approximately four times as far away as Gemini and Apollo satellites, a factor admittedly not conducive to obtaining fine picture detail ("resolution"). The high altitude was selected in order to reduce the drag on the spaceship and enable it to stay in orbit for prolonged flight. Whereas Skylab was scheduled for a particular time period in 1973–74, ERTS-1 would continue to be utilized as long as it remained on course and provided data of reasonable quality. ERTS-1 was still functioning, at reduced efficiency, in late 1974, with ERTS-2 anticipated for the following year.

Skylab Sensors Hold Unknown Potential

Although ERTS resolution has actually been surprising in its clarity, scientists in certain projects nevertheless looked forward to 1973–74 and the occasional Skylab results they might receive. Skylab was to orbit at only 260 statute miles above the earth and presumably would offer better resolution than ERTS. (By and large, it did.)

More important than its resolution, however, was the potential offered by some of its extremely sophisticated sensors. Skylab carried a six-lens (visual and near-IR) camera, two microwave sensors, and a thirteen-channel MSS which included one far-IR band. Since far-IR rays radiate at night, no light source whatsoever was needed; Skylab was a completely nocturnal animal. Even more important, it has the potential of sensing heat emissions from the earth, with such exciting possibilities as locating geothermal power sources.

The unmanned Skylab workshop was put into orbit in May 1973. Three command modules (Skylabs 1, 2, and 3),† each manned by three astronauts, were then successively launched and docked with the workshop, the first crew spending four weeks in orbit, and the subsequent crews eight and twelve weeks respectively. Upon docking of the command modules, the rather grotesque-appearing Skylab cluster was extended to 118 feet of Tinker-Toy-like assemblages. With a windmill array of four solar panels protruding overhead, it resembled an unwieldy sort of helicopter. Directly under the windmill was the multiple docking adapter, a compartment that connected the workshop with the command module. It was here that the sensors—all requiring some degree of manual control—were located and trained on the earth below in assorted sweep patterns. The information collected by the sensors, rather than being sent live to the earth, was carried back by each team of astronauts.

The first Skylab crew reached the workshop facing a major patching chore on the module, and they soon were climbing over the satellite like roof repairmen. It was some days before their attention could be turned to their pre-planned activities in the satellite. And then, since the Skylab mission included numerous events aside from earth observations, competition for power and astronaut time was high. Conditions improved in Skylabs 2 and 3, but even so, sensing time averaged less than an hour a day over the twenty-four weeks of manned flight. There were other limitations inherent in Skylab. Whereas ERTS readings often take months to interpret, the most sophisticated of the Skylab data may take years to program and comprehend fully. Furthermore, Skylab's orbit was not particularly suitable for remote sensing. It

† At first, NASA referred to the workshop as Skylab 1, and to the manned units as Skylabs 2, 3, and 4. We will adhere to the more popular terminology and call the manned flights Skylabs 1, 2, and 3.

spiraled around the earth's midsection on a course much like that followed by all previous manned missions, tracking over limited portions of the earth in the equatorial and middle latitudes and missing polar areas entirely. The coverage was thus sporadic, compared with ERTS and its inexorable monitoring of the entire globe.

If Skylab's performance was limited, the information from its camera (the only Skylab sensor interpreted to date) shows signs of real significance. Skylab's ability to hit upon occasional unexpected strikes of elusive information was apparently its greatest asset. ERTS was the football team that pounded away throughout the game, and Skylab was the place kicker who came in for brief but glorious moments. As one example, ERTS pictures are resulting in the valuable mapping of millions of acres of forest land in the United States, while the Skylab camera has performed more effectively than ERTS in discovering small concentrations of diseased trees. And once the Skylab far-IR information from its more sophisticated sensors becomes programmed and available, its contribution to the study of forest disease may become even greater. Skylab is still a comparatively unknown quantity, but it will eventually expand the achievements of ERTS-1 and ERTS-2. Together these three satellites will point the way for the Space Shuttle's flights of the 1980s and help us determine the feasibility of permanent remote-sensing satellites.

Highly speculative concepts of the future, however, need not dominate our discussion of earth observations. There are such a large number of exciting "today" discoveries which have begun to reach fruition with ERTS that they are the matters our discussion can emphasize, now, in the mid-1970s.

The success of ERTS has mounted quickly and recently. In July 1972, as the scientific world sat poised awaiting the flow of ERTS data, there were mixed moods in the scientific community. Pessimists were looking on the 1970s as nothing more than a time of orientation, of learning a few ground rules which might someday prove useful. Optimists, on the other hand, were looking at the progress which sensing had made from the age of Adam to the age of Apollo, and regarding it as minuscule compared to the mind-boggling advances which they believed lay immediately ahead.

But while diverse opinions as to ERTS's potential success existed, scientists were almost unanimous in their feeling that ERTS was the step which had to be taken, as man sought to determine if the earth could truly be made to grow.

ERTS, in its first two years, has begun to provide an answer.

Part II

THE PROGRESS

4. The Devil Sisters

When Magellan commanded history's first round-the-world voyage, the most daring portion of his journey was the segment that took him westward from South America to Micronesia (1520–21). On that leg, the Portuguese navigator found an immense ocean so calm that he called it "*pacífico*"—a designation, however, which he would surely have discarded had he been able to see beneath the quiet, blue surface waters. The ocean bottom over which he sailed was a thin, weak crust, shattered by frequent earthquakes and punctured by protruding undersea volcanoes. Mercifully, none of these phenomena disrupted Magellan's voyage, and he departed from the world's largest ocean at the Philippines blissfully unaware of the dangers from the inner earth.

Twentieth-century man in the pre-satellite era achieved an awareness of earthquakes and volcanoes only slightly greater than that of Magellan. If a traveler today were to resume Magellan's circling of the Pacific, he would fly north from the Philippines to the rim of China, Japan, Siberia, to Alaska, and finally southward along the west coast of the Americas. From the first leg of such a journey, it would be a spectacular flight over an ongoing combination of gaping canyons and symmetrical peaks. And yet, rather like Magellan, the modern traveler would probably not realize that the canyons were extensively laced with earthquake faults. Nor that hundreds of the mountains he saw were volcanoes—"a ring of fire surrounding the Pacific," as geophysicist Dr. J. T. Eaton describes it.

When the twentieth-century traveler reached North America

and looked down at the coastline, he would see the city of Anchorage, which had been caught unaware by the tremendous 1964 earthquake with its magnitude of 8.5. That quake had made Anchorage into a testimonial to man's inability to forecast even the most extreme earthquake disaster.

Half a continent farther south, in the state of Washington, Mounts Baker, St. Helena, and especially Rainier would loom above the clouds for the traveler to view in their stately white splendor, as calm-appearing as Anchorage and as unpredictable. Rainier would reveal to our observer none of the volcanic tendencies attributed to it in a *Science Digest* article, "The Live Time Bomb in Seattle's Back Yard."[1] (While our traveler would recognize Rainier only as a magnificent mountain, volcano authorities have gone one step further, saying Rainier "will erupt, perhaps this afternoon, perhaps a century from now . . ." This very frustrating degree in inexactitude would seem to sum up what is presently man's highest level of volcanic knowledge.)

Continuing on, today's air traveler could see California's active Lassen Peak, whose frequently bubbling lava befuddles man more than it informs him, and then Mexico, where two major earthquakes occurred in 1973.

Beyond Mexico lies Central America, the strand of land so narrow that only its high mountains would prevent our airplane passenger on the Pacific coast from looking completely across to the Atlantic. Here between North and South America, the threats of the inner earth become at least more apparent, if not better understood. The tremendous mountain ranges of the continents to north and south converge into a Central American pathway, tumbling together like a cord knotted at the waist of the hemisphere. The volcanoes which are squeezed one against another seem obliged to exhale constantly, first one and then another erupting in some fashion, their emissions often visible to the traveler from a distance. On my own first flights over Central America in the early 1960s, it was Mount Irazú which for two years provided the spectacle, constantly showering ash on nearby San José, Costa Rica. A few years later it was Guatemala's Mount Pacaya which regularly belched gases and occasionally threw out lava.

But on the last week of December 1972, Nicaragua's San Cristóbal was the volcano which was apparent from the air. A wisp of gray cloud was spiraling up from San Cristóbal, visible to the passengers of a plane with NASA markings which for a while approached the mountain, and then turned away, southeastward toward Managua.

However, it was not San Cristóbal which at that moment was occupying the attention of Dr. Robert Brown and other U. S. Geological Survey people who were aboard the plane. It was an earthquake, the devil sister of the volcano, which had brought them, a geological team, to Nicaragua from Menlo Park, California. It had been only a week since the disastrous December 23 quake in Managua, and today Brown was coordinating sensing of the city in order to develop a base map—a map which he would use when he was back on the ground, to locate the cracks in the earth that now ran throughout Managua like a spider's web. By finding all the cracks in the streets before they were repaired, he could map the entire web of fractures and determine the actual zones of earth movement.

Brown's map of the actual fractures in the city could then be compared with other information. For example, data was to be derived from a series of seismographs now being installed around Managua by Brown's colleague, Dr. Peter Ward. The seismographs would detect activity deeper within the ground; by monitoring the aftershocks yet to occur under the city, they would add to information concerning the location of faults and epicenters of tremors.

The pictures obtained by the NASA aircraft, however, would be the principal source of information for the map. (For an area as small as Managua, they were of a more usable scale than ERTS pictures which had been taken the week before.) Later, when the Nicaraguans would rebuild Managua, they could avoid putting buildings along all the zones which Brown's group would designate, instead covering the lines of the spider's web with wide green parkways. At least these highly earthquake-prone areas should not again have the opportunity to bury people beneath the rubble of buildings.

This was not a sophisticated project, but it was a first step in applying the technology of remote sensing to the limiting of earthquake disasters. Although man had not yet learned to forecast earthquakes, he could at least better prepare himself for them.

ERTS—Stethoscope for Volcanoes

The Managua project was an example of solid, if not spectacular progress. But the USGS intended to turn space science upon Nicaragua in more ambitious projects as well. The smoking mountain that had wafted its signal outside the NASA plane of Robert Brown was soon to be the scene of another American project, a pilot study that would utilize ERTS as a super-stethoscope monitoring the signs of life inside San Cristóbal.

Within those first weeks of 1973, Dr. Peter Ward had selected a series of sensitive points on San Cristóbal. At each point he installed a device to detect volcanic activity and carry it by landline down the slope to an unmanned information collection "platform," a small box with a dome-style antenna. The box housed a Multi-Level Event Counter (MLEC), a data system which sorted out the signals from the volcano and used the antenna to relay them to ERTS. By February the system was in operation, and it has functioned on a daily basis ever since.

The information being handled is of two types. A constant stream of seismic data, capable of recording the multitude of very small earthquakes which occur in volcanoes on the verge of eruption, flows out to the MLEC from one of Ward's network of recorders. A second network collects signals measuring volcano "tilt," indicating whenever the sides of the volcano are swelling in response to lava pushing up from the depths of the earth. Together the seismic and tilt data create a profile indicating the probability and the timing of a volcanic eruption. Once

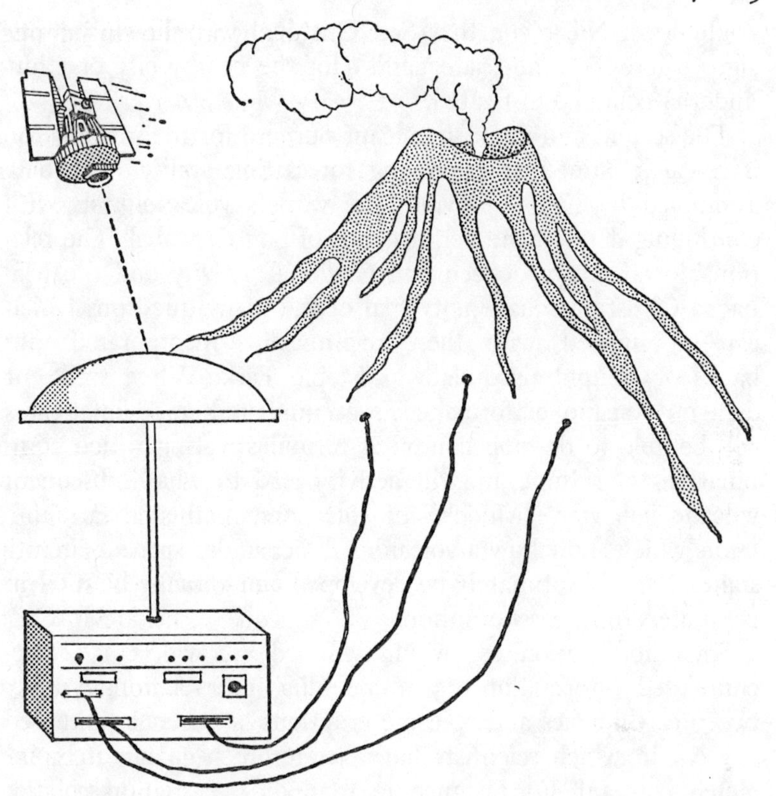

Figure 4: Volcanic information gathered at the domed platform is telemetered to ERTS and relayed to U.S. earthquake center within ninety minutes.

the signals reach the MLEC they are automatically compressed into an abbreviated form and then telemetered to ERTS. Utilizing its Data Collection System (DCS), ERTS functions as a communications relay, retransmitting the information to Menlo Park for human observation.

The observations are keyed to two purposes, the most immediate being operational use. The data is monitored constantly in Menlo Park to detect profiles suggesting any approaching volcanic eruption (within the limits of today's primitive forecasting technology). If the situation warranted, Menlo Park could warn Nicaragua that San Cristóbal was showing strong signs of erupting, and safeguards (for the nearby city of Chinandega) could be instigated.

The second and more significant purpose for the information is research, aimed at improving forecasting ability. The data assembled by ERTS provides the world's volcanologists with continuing data on the performance of San Cristóbal. The relationship (if any) between San Cristóbal activity and the Managua Christmas quake is typical of the many questions which can be pursued, with these relationships incorporated into banks of computerized data at Menlo Park. When sufficient data on volcano performance is accumulated, mathematicians will be able to develop computer formulas stating which combinations of seismic and tilt activity lead to what subsequent volcano behavior. "Models" of volcanoes, mathematical simulators which show how a volcano will act under specific circumstances, should ultimately be developed and actually be used in computers to forecast eruptions.

Such modeled forecasts would be based on a wide range of accumulated information—again including interrelationships between earthquakes and volcanic eruptions, a generally accepted concept in which scientists have heretofore been almost completely ignorant. For instance, in 1972, did great chunks of the earth's plate, on the floor of the Pacific Ocean, buckle and drive downward against the Central American shoreline, puncturing the earth's mantle, and cracking open earthquake faults in Managua? And then in 1973 while the earth was still quaking, could hot materials be released from the mantle, a hundred miles down—to shoot up through crustal conduits and out of the mouth of San Cristóbal or some other volcano? Or was another, larger earthquake likely to occur in 1973, perhaps a thousand miles north in Mexico? (As it turned out, both phenomena occurred.) By the early 1980s, satellites will perhaps have collected the means of understanding these violent relationships, along with some degree of ability to forecast both tremors and eruptions.

While speaking optimistically of the future, we should note

that there is nothing startlingly new about either of the monitoring devices used on San Cristóbal. Tilt meters have been used for decades, and seismographs for centuries, although until now their recordings have been interpreted by scientists who were stationed at the volcano sites themselves. A permanent observatory in Hawaii has monitored Kilauea and Mauna Loa since 1912, with a staff of dozens of scientists and assistants on duty. Obviously this kind of local observatory can be quite elaborate in its data collection. However, the continually accumulating expense of operating the Hawaii observatory has run into many millions of dollars, and its chances of detecting activity are limited to a single volcanic area. The expense involved in an observatory has precluded establishing more than a handful of such installations in the world, and volcano science has made proportionately limited progress.

Obviously, man could advance the science of volcanology considerably more if he were to expand his study. There are many hundreds of volcanoes on the earth which have some likelihood of erupting; a widespread monitoring of all of them would not only advance the development of models, but would also allow warnings of eruptions, avoiding tremendous losses of life and property.

It was with this concept in mind that geophysicists and geologists at Menlo Park established the ERTS "thin surveillance" system, of which the MLEC installed by Peter Ward at San Cristóbal is a part. The thin surveillance system is a pilot program of twenty MLEC platforms, each collecting a thin level of information from a single volcano (or a small cluster of volcanoes) in the Western Hemisphere. Each MLEC, like the one at San Cristóbal, telemeters abbreviated volcanic data to the United States via ERTS. Within an hour and a half of the time the data is first telemetered from the volcano area, the information is available in printed form at Menlo Park, where all the input from the various sites can be effectively and economically studied.

Shortly after the ERTS launch, the first MLECs had been installed in California, Washington, and Alaska in a dozen volcanoes, three of which had been active in the twentieth century. Then during the month following the Managua quake, MLECs were installed, not only at San Cristóbal, but in six other volcanoes in Nicaragua, El Salvador, and Guatemala. These Central American units thus completed a chain of nineteen critical surveillance points running from Nicaragua to Alaska. Together they would monitor a 400-mile segment of the volcanic "ring of fire" surrounding the Pacific.

The Value of ERTS Surveillance

The merit in monitoring not one but many volcanoes became quickly apparent. On February 13, 1973, an MLEC was installed at the active volcano Fuego in southern Guatemala. On February 19, the seismic recordings being collected by the MLEC increased markedly in strength and number, passing what was later recognized as the threshold of pre-volcanic tremors. On February 22, the volcano commenced an eruption of ash which showered down on the surrounding farmlands for a week. All this while, San Cristóbal, then the most feared volcano of Central America, exhibited a surprising tranquillity (a mood which it maintained throughout all of 1973). If scientists had been able to have monitored only one volcano in Central America in 1973, they would undoubtedly have chosen the wrong one—San Cristóbal. And Menlo Park would not have obtained the Fuego pre-eruption data, which is now serving as a useful guide for forecasting future eruptions in Central America and elsewhere.

Another example of the need for extensive surveillance had been brought into focus a month before the Fuego eruption. It had occurred not in Central or North America, but in Iceland. Just outside Reykjavik, Iceland, were two volcanoes, chosen as the site for the only MLEC platform to be placed outside the Pacific ring. The platform was installed in December 1972. On January 23, 1973, Helgafell, another Icelandic volcano, located on an island ten miles off the coast, had erupted unexpectedly and turned the nearby village of Vestmannaeyjar into a raging inferno. (On the mainland more than fifty miles from the volcano, the newly installed MLEC recorded no hint of what was occurring. Effective for a radius of only a few miles, the unit's purpose had quite properly been to monitor only the two volcanoes adjacent to it.)

It is significant that Helgafell was believed to have been inactive for fully seven thousand years, and yet now she had explosively awakened. Meanwhile, scores of active volcanoes around the world—including ten in Iceland itself—remained completely quiet. We note that until Helgafell had come violently to life, it had been considered a majestic peak, and nothing more. Authorities had assigned to it none of the potential attributed to Rainier, for example. In fact, said Ward, "Helgafell would not even have ranked among the world's five hundred most-likely-to-erupt volcanoes."

Considering Helgafell's unfearsome reputation, it is obvious that prior to 1973 no one would ever have suggested the installing and staffing of a complete observatory on its slopes. "It is hard to get support for monitoring [volcanoes] when eruptions can be expected so rarely," said an article in a 1970 science journal.[2] "But adequate monitoring systems could save thousands of lives by warning of eruptions that are bound to occur —sooner or later."

The ERTS surveillance package, which costs only two thousand dollars, has the potential of bringing this monitoring within the financial reach of virtually any community in the world. If the present ERTS experiment with its twenty units proves successful, hundreds or even thousands of volcanoes might be considered for future monitoring, as part of an inexpensive international function, carried out on either a paid or aid basis.

The collection of data from a broad network of volcanoes would not only provide a worldwide warning system, but would also have tremendous research benefits for the United States itself. By including more volcanoes in the surveillance, the opportunity of monitoring eruption would increase, and along with that the development of models for particular volcano types. For example, San Cristóbal is much the same volcano type as Washington's Rainier, California's Shasta, and, for that matter, the infamous Krakatau. (Krakatau, like Helgafell, lay isolated on an offshore island, but nevertheless provided an airborne rock barrage that killed 15,000 people within a fifty-mile radius on Sumatra and Java in 1883.) By developing models from performance histories of numerous Krakatau-type volcanoes and utilizing the models to continuously monitor all similar volcanoes, invaluable forecasting capabilities could be developed.

Such forecasting expertise would be of considerable comfort to the hundreds of thousands of people living near Rainier, that "live time bomb" capable of erupting at any time. Once a Krakatau-type model could be developed and put into operation, the possibility of a Rainier eruption coming unannounced would be very remote. Instead, Menlo Park should be able to give residents of the state of Washington a series of progressive warnings, and evacuations and other emergency preparations could all be commenced well in advance.

Heartbeats from Active Faults

The volcano is perhaps less understood by scientists than is her sister devil, the earthquake, the latter having been much more extensively documented, in recent as well as historical times. For example, a variation of the thin surveillance system we have been discussing for volcanoes has existed for earthquakes since 1971; it has, however, not involved satellites. Some one hundred points in and around San Francisco are being constantly monitored by seismographs, with the information transmitted to Menlo Park by landline, rather than by ERTS. Seismic recorders, in critical earthquake faults, provide scientists in the Menlo Park earthquake center with a constant awareness of fault movements in and around San Francisco. The information is primarily of use in research at this point, but to some degree it is being relied upon operationally.

Its operational potential came dramatically to the public's attention in March 1973 when the earthquake center issued what was perhaps the United States' first responsible earthquake forecast. Based on seismic data recorded along the San Andreas fault southeast of Hollister, California, the earthquake center predicted a fault movement of 4.5 "within the next few months." Fortunately (or unfortunately, from a pristine forecaster's point of view), the quake did not occur.

Regardless of the failure of this first forecast, the very fact that this surveillance information enjoys such a degree of confidence among earthquake scientists at the center is promising in itself. The use of widespread seismic monitoring will now be watched with tremendous interest everywhere. As forecasting success does develop as it almost surely will, more and more earthquake-conscious areas will want to be included in the fault monitoring—a seemingly excellent idea, although fraught with enormous logistics problems. While the landlines have proved convenient as a means of monitoring the San Francisco area, they would scarcely prove adequate to monitor all of California, let alone the entire United States or the world. The idea of a labyrinth of cables running from all over the globe to Menlo Park does, needless to say, seem impractical. Furthermore, any such landlines would be disrupted when earthquakes occurred and be immediately rendered useless at the precise moment when they were needed most—in the midst of transmissions critical to locating and measuring the quakes.

Microwaves, to be transmitted along the earth's surface just as television signals are today, were at one point considered as an alternate means of carrying the seismic data. However, the cost of microwave relay stations necessary to link just one distant earthquake-prone area, such as Alaska, to Menlo Park could run into hundreds of thousands of dollars. Therefore, believes Eaton, "satellites definitely offer the most practical vehicle for transmitting earthquake information, just as they do for volcano data." Inexpensive seismographs along all active fault lines, regularly monitored probably by one of the weather satellites (see Chapter 11), will, Eaton hopes, eventually be installed. As a result of this and other research, earthquakes (probably sooner and more reliably than volcanoes) stand a good chance of being centrally observed and forecast on a regular basis.

ERTS's Remarkable Contribution to Forecasting

There are quite a number of factors which caused Eaton in 1973 to anticipate man's ability, within ten years, to make earthquake forecasts "as reliable and precise as hurricane warnings." Several of these factors are not satellite-related, and we need not discuss them here. However, a major contribution has recently been made by ERTS pictures, seeming to provide a new basis for all earthquake study; ERTS pictures have already more than doubled the number of faults known to man and available for earthquake observation.

There are two scientists whose work seems particularly to have related to this achievement. Dr. Monem Abdel-Gawad and Dr. Joel Silverstein, staff scientists at Rockwell International Science Center in Thousand Oaks, California, have both been faithful fault watchers since the early pictures from space. During the Gemini and Apollo era, they, like numerous other scientists, were impressed with the abundance of faults which began to appear in the space photographs. Fault lines which had illusively appeared and disappeared for the geologist hiking on the ground, were now being clearly recognized from space.

Abdel-Gawad and Silverstein found they not only could map the faults on space pictures, but often could determine that a fault had recently been active, by noting that the land on one side of it had moved laterally. If a space picture showed that a man-made road had been ripped apart and offset, it was likely the scene of a twentieth-century quake. Similarly, if a stream jogged across a fault, it was probably the result of a quake which had occurred at least within "modern" geological times—the last million years—therefore indicating the fault to be still potentially active and dangerous.

Among those areas which had been photographed by Apollo

were scattered rectangles in southern California. Abdel-Gawad and Silverstein became interested in the California patterns of transverse (east-west) faults, which intersected the infamous San Andreas lateral (north-south) fault. Silverstein suspected that "stresses could be built up at fault intersections, where the San Andreas could transfer its pressure to a transverse fault."

Up until then, the large lateral faults had occupied everyone's attention, with but scant notice taken of the transverse. The idea that intersections might be the source of quake-producing pressures had, however, been advanced in 1965 by Dr. C. R. Allen of California Institute of Technology, and now Abdel-Gawad and Silverstein began using satellite photography to pursue the theory. Numerous fault intersections appeared distinctly on the new space photos, and the scientists asked themselves if these intersections were probable areas for future earthquakes. To pursue this question, the scientists assembled thirty years of earthquake history. On an overlay of the Apollo photo, they plotted the recorded epicenters of each earthquake from that time period, in an effort to see if the earthquakes had occurred at the newly discovered fault intersections. As it turned out, some correlation was noticed, but the scientists found more questions than answers as they moved on in the study; nothing was definitive.

Then came February 1971, when an earthquake of 6.6 magnitude savagely ripped open the San Fernando Valley twenty-five miles north of Los Angeles. The quake occurred not on the San Andreas, as first assumed by the public, but along a transverse fault zone which intersected the San Andreas, providing a perfect example of the concept Abdel-Gawad and Silverstein had been exploring.

The quake had occurred in a fault zone that had been identified for some time as the Sierra Madre zone, but the individual faults which caused the quake had gone completely unnoticed in the past. They were now quickly identified and named—Tujunga fault for the disrupted Tujunga Canyon, Sylmar fault for the hard-hit city of Sylmar, and Veterans fault, which ran under the VA hospital where fifty lives had been taken.

Abdel-Gawad and Silverstein were now being given an opportunity unparalleled in geological history—the opportunity actually to see over-all change in an area which they were currently studying. In eighteen months, ERTS would provide a post-earthquake view of southern California; by comparing ERTS with Apollo pictures, the researchers could see how the earthquake had displaced the faults, and they could begin working toward an understanding of what might come next.

The two scientists asked for ERTS data covering southern California and millions of square miles in other earthquake areas, and for additional Skylab pictures of particularly critical fault intersections. As soon as the ERTS imagery began to be widely distributed, it became apparent that the faults previously discovered by the scattered Gemini and Apollo photographs had barely scratched the surface—not only because they covered such small random portions of the earth's surface, but also because many of them were taken from oblique angles and often failed to capture the fault lines. ERTS, by scanning at right angles to the earth to minimize distortion, and by providing a continuous mosaic, revealed a wealth of faults which had not been noticed in Apollo pictures; the earth was obviously cracked and split open in more places than anyone had ever realized! For the first few ERTS pictures, fault locating became the major pastime of the remote-sensing community. Scientists everywhere, often having no particular interest in earthquakes, but having access to various ERTS images, began to report their discovery of probable faults and intersections in the particular pictures they were studying. The active faults (those showing evidence of "recent" movement) were of course of particular interest to Abdel-Gawad and Silverstein.

By now the work which Abdel-Gawad and Silverstein had previously done—manually plotting historic earthquakes and comparing them with fault intersections—had been upgraded at the Rockwell Center into a computer mapping system. The two scientists had already fed the locations of historical earthquakes into the computer, which now electronically compared the quake locations with the increasing number of recognizable intersections. The new active faults being discovered by ERTS were also added to the computer input. As they studied the computer's findings, Abdel-Gawad and Silverstein made two observations: First, areas where earthquakes had been known to occur were almost invariably characterized by the presence of active faults. Second, the reverse was not necessarily true; that is, there were many active fault areas where absolutely no earthquakes had occurred in the last thirty years.

The explanation seemed obvious. Up until now, the only areas where man had expected earthquakes were those places which had quaked since accurate records had been kept—usually only in this century, a mere wink of an eye in the time span of geology. By being alert only to those areas of twentieth-century activity, man had been constantly surprised by tremors occurring in new places. The 1971 San Fernando quake, for example, came in an area where no previous earthquakes had ever been recorded. (So naïve was man that he had built the Veterans hospital virtually astraddle an active fault.) By basing his expectations on recent experience, man had in fact built an entire civilization with disregard to a tremendous number of dangerous faults. Satellites offered man an opportunity to become aware of these previously ignored areas. "Now, with ERTS, we can identify these danger zones and at least begin to take notice of them in our future planning," said Silverstein. Instead of considering only a few decades of earthquake history, man could begin to base his decisions on the record of a million years.

On March 14, 1973, Abdel-Gawad explained the new-found danger spots in a television interview. As an example, he pointed to one particular fault which ironically ran past the science center where he worked, and near his own house in Thousand Oaks. "This," he announced, "is an active fault."

The response in fashionable Thousand Oaks was one of alarm. Abdel-Gawad tried to play down any hysteria by pointing out that "I live and work in this town myself." (He prudently avoided mentioning that he was, perhaps coincidentally, in the process of moving to another neighborhood the same day.)

The following day, Silverstein explained the matter this way:

"That one fault was just a random example. California has dozens of equally dangerous faults that we have discovered through ERTS. And there are several previously known faults which ERTS has now recognized as recently active. But each of these has only a very small mathematical chance of becoming an earthquake epicenter.

"What we suggest is not that we evacuate Thousand Oaks, but that we use the new knowledge of faults for future planning —and that we take a good look at locations of schools, hospitals *—and, especially, nuclear plants.*" The immediate application of ERTS data, for zoning purposes, obviously had immeasurable potential value.

Silverstein did not mention dams as one of the installations for which hazardous sites should be avoided. However, several years before, in studying an Apollo photo of Hoover Dam, Abdel-Gawad had detected a previously unknown fault. He had then traveled to the dam to check on it firsthand. He found it—right at the edge of the Hoover Dam parking lot. Since it was well known that small quakes had occurred elsewhere in the canyon ever since the dam's construction, this fault's close proximity to the dam was alarming. And while it is obviously preferable to learn about faulting *before* a dam or other structure is built, there are engineering safeguards which can and should be taken at places like Hoover, once the problem is recognized. The ability to locate active areas and determine varying shades of earthquake probability is likely to be developed rapidly, as more and more data is sensed and analyzed.

In February 1973 an earthquake occurred near Oxnard, California, in a virgin epicenter location, providing an excellent opportunity for geologists to study the "before" and "after" appearance in two successive ERTS pictures. This type of opportunity will continue to occur, and within the next few years, Abdel-Gawad and Silverstein may make it possible for earthquake science to develop computerized earthquake models which were inconceivable prior to ERTS. As the two scientists assimilate more and more pictures of geological changes, and incorporate them in their computer, they will develop new geometric concepts. Intersections, fault activities, and earthquake history will produce geometrical patterns that will be an extremely useful factor in forecasting the location, although probably not the time, of future earthquakes.

Laser Ranging for Long-Range Forecasting

Time-forecasting of earthquakes may someday be achieved through one satellite concept which relies on the use of laser ranging stations. Two ranging stations, located in San Diego and Quincy, California, are 550 miles apart, and on opposite sides of the San Andreas fault. By bouncing laser beams off a satellite, the two stations have the ability to determine their own exact locations on the earth. In so doing, they would also be precisely determining the distance between themselves. The variations in this distance, as time passes, would presumably be a result of gradual movement by the San Andreas, as a result of a strain accumulation. Thus, the movement of the San Andreas could apparently be periodically measured with laser precision.

The U. S. Geological Survey, in a May 1972 report,[3] calculates that if the rate of San Andreas movement proved to be two centimeters per year, an earthquake on that fault the size of the 1906 San Francisco quake (8.3 magnitude) should not occur for the next four centuries. If the fault movement is found to be greater, the time until the predicted earthquake date will be sharply reduced. For example, if the rate of movement is measured at six centimeters per year, the quake would be expected within fifty years.

This extraordinary long-range-forecasting technique, when and if fully developed, will owe its existence as much to satellites as to the laser technology itself. (The USGS report also explores the possibility of making the same measurements with a land-laser system, instead of with the satellite-laser combination. The report concludes, however, that in order to duplicate the satellite system's accuracy, it would be necessary to establish an unwieldy complex of thousands of landlines and numerous control stations throughout California. "Satellites will have to be the name of the game," commented one USGS scientist.)

The extent to which laser ranging can be applied is still a matter of conjecture, but the system does seem to offer high potential in relation to such major faults as the San Andreas, thus adding another satellite system to help forecast the assorted vibrations and explosions of the earth.

Explosions—Isolated or Interrelated

In the middle of the violent month of January 1973, even before Helgafell turned its fiery furnace loose on Vestmannaeyjar, Guatemala's Pacaya expanded its flow of lava into a recognizable eruption. And ten thousand miles to the west, on the opposite side of the Pacific's great ring of fire, a volcano erupted in Japan on February 1. An earthquake shook China five days later.

Did the earthquake in Managua affect the eruptions in Iceland or Japan in any way? Were Fuego and Pacaya isolated incidents or were they interrelated actions? Few scientists are yet prepared to say. The earth is constantly cracking and exploding, but relationships are still not recognized. The development of computerized history of these activities, say some scientists, needs to be commenced intensively. Geologist Ira Bechtold believes in continuously collecting data on strain and crustal magnetism from points all over the globe. "The plates that make up the earth's crust are virtually sliding around on the mantle beneath them," says Bechtold, who finds the possibilities of interaction many and exciting. Measurements of motion in the Eastern Hemisphere, Bechtold has conjectured, might someday lead to earthquake or volcanic forecasts for North and South America, and vice versa.

In spite of this optimism, man in the mid-1970s stood with all the proof ahead of him—with few axioms, assorted schools of thought, and many controversies existing. On the last day of January 1973, Dr. Peter Ward sat in the Menlo Park center and remarked to me, "We don't really even know if earthquakes and volcanoes are related."

The devil sisters would seem to have been listening; less than two hours later, Ward was handed a teletype from Mexico reporting one of the earthquakes of the century (7.8), with unofficial reports claiming the eruption of a nearby volcano.

The epicenter of that earthquake was fortunately in a sparsely populated area of western Mexico, with only ten deaths occurring—although a full three hundred miles to the east, the skyscrapers of Mexico City teetered threateningly, an appropriate ending for the convulsive month of January. It had been a month that would be remembered not only in Mexico and nearby Guatemala, but far to the north in subarctic Iceland, just as February would be memorable from Guatemala's Fuego to China and Japan. The plates of the earth were sliding on all faces of the planet, continuously colliding one with another, puncturing within and exploding without, a plethora of turbulence occurring in patterns still imperceptible to man.

We would surely require a synoptic view to comprehend it.

5. The Energy and the Metals

"Gold. Dig here."

A neon sign flashing those words was what people seemed to expect to find in satellite pictures, thought geologist Ira Bechtold. "Copper straight ahead. Oil to your right." The public, or at least those few who were aware of remote sensing at all, had been injected with some utopian ideas about what could be done to locate minerals underground. Still, Bechtold said to himself, if earth observations could not offer neon, they could at least raise some subtle "stop" and "go" signs for the person who knew the language of sensing. And the signs, faint though they might be, could be lighting the avenues to man's whole range of future sources of energy and mineral wealth.

Today Bechtold would be talking to a group which, although relatively few of them were geologists, would respond to this new language of geology. Most of his audience would be space applications people, Charles Matthews among them. They were gathered to hear Bechtold and others on the opening morning of the American Institute of Aeronautics and Astronautics conference, at the Sheraton Park Hotel in Washington, D.C.

It was January 8, 1973, and aside from the final Vietnam negotiations taking place in Paris that day, the most-discussed topic in the U.S. capital was the energy crisis. Two years before, the President had assigned an array of task forces, each to review one or more of the world's potential energy sources, to see how they might fit into the decades and centuries ahead. The sources assigned for the study had been wide-ranging: the anciently utilized wind and water, the dwindling gas and oil, the once-disdained but now newly respected coal, the emerging but controversial nuclear, and the exotic tidal, solar, and geothermal possibilities. It was a long-range program, and on this day in 1973, relatively little had been firmly established, yet suddenly the energy crisis was rising to the forefront of concern. Congress was milling about restlessly, with half a dozen major energy-related committees, gathering momentum and with lawmakers realizing that Spaceship Earth was running out of fuel.

The energy crisis had come upon the scene so quickly and unexpectedly that few of the ship's passengers were taking it seriously in January 1973. But projections from industry on increased power needs were being punctuated with urgency, and the future seemed black, or at the very least brown. In the short run, the Denver schools would close for lack of heat before the month was out, and brownouts and mandatory cutbacks would arise throughout the nation within the year.

Whether or not one was suspicious of the factors which had created the immediate crisis, however, no one could doubt the severity of the fossil-fuel shortage in the long run. The Club of Rome had said that natural-gas reserves would fail to meet worldwide demands by 2029, with petroleum reserves reaching the same point only one year later.[1] (Futhermore, the Club had based its unnerving estimates on the fairly optimistic assumption that man would actually unearth five times the present known gas and oil reserves. Other, equally reliable studies soon to be released would be even more discouraging.) It was on these critical areas that Ira Bechtold's talk on geology today would particularly impinge.

Synoptic and Super-Synoptic

"Here's what the earth looks like from a spaceship a tenth of the way to the moon." Bechtold flashed what he considered a spectacular color slide on the screen. It had been taken just the month before by the outbound Apollo 17 astronauts 25,000 miles "southeast" of the earth, with the center of focus on South

Africa. It was an unusual photograph of the world in that it offered a cloud-free view of Antarctica, and even more so because the full disk of the world was illuminated—if moonmen had existed, and had seen such a planet beaming down from a balmy evening sky, they would conceivably have waxed poetic about a "full earth."

As for his audience of earthlings here in the Sheraton, observed Bechtold, they were rather complacent. Five years before they would have rendered a full-strength "gee whiz" gasp at such a picture, but today he was content to hear them mildly murmur appreciation for its color and resolution. It was just as well, he thought. He wanted their attention to be, not starry-eyed, but discriminating and critical as he showed them evidence of such things as continental drift; the drift theory, which ten years before had been held by a handful of "eccentric" scientists, had been bolstered dramatically by space imagery. Today, on Bechtold's picture of the full globe, continents appeared strikingly adrift, in a way that no two-dimensional map could have duplicated. The Red Sea rift between Arabia and Africa seemed to "stretch on the globe like a wrinkle on your knuckle," Bechtold said.

He flashed a closer picture, taken 15,000 miles from earth, this one of the Western Hemisphere, with the southwestern United States the focal point. "Here you can see the San Andreas fault in a way that no earth-orbital flight could reveal it." Just as ERTS could replace thousands of airplane pictures, this one more distant shot could encompass thousands of ERTS images. It was a *super*-synoptic view.

"We may change our ideas about the San Andreas fault. We've always thought it went down here into the Gulf of California." He pointed at the screen. "But here we see another line running farther over to the east, into Mexico.

"Which is the fault? Or are there two of them?" Bechtold's question was one which geophysicists would be striving to answer.

Now he flashed on another Apollo picture, this time one taken from 10,000 miles, and then yet another from 6,000 miles, each centered on the southwestern United States. Bechtold drew his audience's attention to the Hoover Dam area. Then he showed an image taken from hundreds, rather than thousands of miles away, by a Nimbus satellite, circa 1967. The picture covered an area less than 200 miles wide centered on Hoover Dam. Lakes in the Sierras and other features absent from the Apollo pictures now appeared (the size of pin pricks), and fault lines which had been indistinguishable at greater distance now stood out. Next Bechtold showed an ERTS picture of the same area, closer and with markedly sharper resolution. "As we get closer and closer, you can see that the dam and these fault lines are identical from one picture to the next, even though they were taken years apart with completely different sensors. You can," he added, "be satisfied there is no optical illusion."

The picture on the screen was one which Bechtold's two young colleagues, Mark Liggett and John Childs, had used to locate "dike swarms" on either side of the Colorado River near Hoover Dam. The swarms of dikes were fissures that had been filled with molten material ages ago. They extended along the river for fifty miles and included an area where there were several active gold mines.

The dikes were so immense and so close to Hoover Dam it seemed amazing that they remained unknown till now. "On the ground, you can't see the mountains for the rocks," Ira often philosophized.

The knowledge of the fissures could be useful on the ground, however—extremely useful to the gold miners working that area. If a mining company noted, from this satellite information, that its present successful mine lay on a particular dike, then it might be deduced that very possibly there were other beds of ore lying elsewhere in that same dike, or in an adjacent one.

Bechtold was moving along now in his presentation, showing pictures of other areas. "As we see things which require closer examination, we fly over critical areas with the U-2 aircraft." He brought on a picture taken at 70,000 feet.

"Or lower." He flashed on an image obtained at 10,000 feet, by an airborne radar. A single fracture zone now yawned from the screen where the global views had been a few moments before. In that short time the audience had seen the earth come toward them as though they were viewing it with a zoom lens. They had started looking at the entire globe, and it had moved closer and closer toward them, showing smaller and smaller segments of itself until now a canyon gaped in front of them on the screen. "I feel as though we'd been pulled magnetically in from space to a rim of the fault," someone in the audience commented.

"Zooming In" for Minerals

Surveying the world from many altitudes was an intriguing way for any geologist to look at the earth, whether he was engaged in a study of the earth's history, or in earthquake research, or in a search for minerals and oil. And it was a technique that was arriving in the nick of time. For years, miners had surveyed the world in search of minerals that were in outcrops of rock on the surface of the ground, and indeed virtually *had* cried out, "Gold. Dig here."

It had been great while it lasted. But in the last few years, the world has exhausted most of these easy conquests. Treasures have become harder to find and, to compound the felony, lower in quality. Discoveries of high-grade oil and ore have been reduced to a point where lower and lower grades of fossil fuel and minerals are being brought out of the ground.

"We're tickled pink with bodies of ore that we wouldn't have bothered with at one time," one mining-company engineer puts it. "And even those are harder to find." In the case of minerals, especially, geologists are having to look for more subtle, indirect means of determining where the ore lies, finding geological structures which are not valuable in themselves but which simply hint at hidden mineral locations down deep below.

Geologists have a wide range of ideas about how much wealth still exists in the ground. The Club of Rome's method of estimating mineral wealth, in which each ore is arbitrarily assumed to total five times its *known* reserves, very well might be overly optimistic for some minerals. But, on the other hand, the sweeping new style of search with space pictures, viewing the whole world first and then zooming in on detail, might reveal that the club's estimates of mineral inventory are actually conservative. Extensive reserves almost surely lie in portions of the earth's crust where miners have never previously had occasion to look. A spot in the wilderness that appears innocuous in itself might show up in a space picture as part of a rich-appearing rock structure running a length of ten miles, or of five hundred miles. Or of the entire globe.

"Up till now, we just haven't been thinking big enough in geology," Bechtold concluded.

The Red Sea and Black Gold

In the early days of sensing, North Africa and the Middle East provided more clear skies along the routes of the Gemini flights than any other part of the world.

Dr. Monem Abdel-Gawad, who was an Egyptian-born American, found this convenient; he already knew the geology of the Middle East from years of field work along the Red Sea shores. Now he was intrigued with the prospect of studying the same area again, but this time from his Rockwell International Science Center office, by means of a synoptic picture.

He obtained all the Gemini pictures shot by the astronauts with their 70-millimeter cameras and examined them in his laboratory at Rockwell International Science Center. One day in 1966 he was studying a Gemini 11 photograph taken of the Red Sea area when he noticed the faint signs of fault lines along both the northerly and southerly shores of the sea. The patterns interested him, and he spent several hours in his laboratory, examining and re-examining the pictures with magnifying glass and calipers.

Suddenly two of the faults which were trailing off from the north shore of the sea caught his eye. They seemed to resemble closely two faults on the opposite shore. With more than casual attention, he measured and analyzed the two pairs of faults, his interest increasing as he found that the north-shore and south-shore fault lines seemed to coincide perfectly. He measured and remeasured, his excitement growing; he was beginning to suspect that he was on the brink of forging another link in the chain of support for the theory of continental drift.

Geologists had long argued over the possibility of the Red Sea being a giant fissure formed by the separation of Africa and Arabia. For decades, geologists had studied the formation of the sea, the structures of rock formations along its shoreline and the deposits of metals and brines in the sea itself. The result had been a lively controversy over continental drift.

But now, the Gemini pictures were dramatically revealing a jigsaw pattern to Abdel-Gawad that had not been visible from ground or airplane. The northeastern shore seemed to have drifted ninety-five miles northward, he noted. This would mean that the Maqna Block, a triangular piece of land on the north shoreline at the mouth of the Gulf of Aqaba, would have previously lain right in the sea itself. Among other things, this indicated that the Block was an area of high petroleum probability.

Excited as he was over the pictures, Abdel-Gawad displayed the scientist's characteristic restraint and refused to accept them at face value. The Gemini pictures were strong evidence, but not proof. He spent the next two years delving into every major geological paper that had been written on the Red Sea. He compared aerial photographs and ground surveys with his Gemini pictures, attempting to satisfy his scientific urge for more data.

Ultimately, however, he became convinced of his findings, and in 1969 published a paper stating that new evidence had been found to substantiate Arabia's drift northward. He also suggested that "petroleum prospects in the northern part of the Red Sea" would have to be considered.

Later that same year, the exploratory Barqan oil wells were drilled in Maqna Block. There was an immediate flow of natural gas, dramatically confirming Abdel-Gawad's speculation. The two wells were, significantly, exactly where Gemini and Abdel-Gawad had pointed their fingers.[2]

It might have been coincidence. However, drilling operations on both sides of the Red Sea in the next few years will have a chance to reinforce or refute the theory. If new oil discovery areas on the northerly side of the fissure coincide with the already existing areas on the southern side, the evidence will be strong that the two shorelines were once united—and that the Gemini pictures have proved out as tools for oil prospecting.

If favorable, Gemini and the Red Sea will have been just the beginning. Man has an entire globe to search, with pools of black gold buried in an unknown number of treasure spots. Many geologists are convinced that there are vast deposits north of the Prudhoe Bay strikes; but to find them—and develop them safely in the fragile environment—is the challenging problem the geophysicist faces. Oilmen are in much the same position as the household handyman trying to hang a picture on the living-room wall, hopefully on a stud; before drilling numerous holes, whether in the ground or in the wall, it helps to know where to start. If satellite pictures can single out high-probability petroleum areas, as when Abdel-Gawad noted the resemblance between known and unknown oil areas—then the geophysicist can be much more selective in exploratory drilling, pursuing only those areas with high-probability signatures. From the standpoints of both operating thrift and environmental impact, this advantage is recognized by all.

Abdel-Gawad did his study based on the casual Gemini pictures, a most insignificant source when compared with the present accumulation of ERTS data. Oil companies now have access to a complete library of ERTS pictures covering productive petroleum areas, where signatures can be developed for favorable geological conditions—and then extended to searches for new high-probability areas. Furthermore, this same search technique can be expanded tremendously by any company that uses ERTS data in one of the available automated electronic projections.

And use it they will, albeit behind closed doors. The competition for secret caches of wealth has caused oilmen (and miners) to be typically a quiet group, communicating little with the outside world about their plans.

Their comments about their use of ERTS have been no exception. Publicly, their attention to ERTS has been minimal. However, their private interest has been more than healthy. One oil company we know of has purchased ERTS data (in magnetic tape form) *for the entire world;* while other ERTS users have to order pictures from the satellite film library in Sioux Falls, South Dakota, this particular company has provided itself with the capacity to examine high-probability areas on a worldwide basis, at will.

A New North Slope Strike?

Nevertheless, while oilmen remain silent, areas with high petroleum potential sometimes receive public attention as a result of reports made by noncommercial groups, usually the U. S. Geological Survey. The first such report of the modern satellite era was based on a picture taken by ERTS on July 27, 1972, just four days after its launch, and reported in the *Oil & Gas Journal* of May 28, 1973.[8] The picture covered a government land area in the Alaskan coastal plains, south and west of Prudhoe Bay. At the eastern edge of the picture were the Umiat oil fields, which had been developed to meet the U. S. Navy's petroleum needs in the 1940s and 1950s. In the northern portion of the picture was a large lake-filled plains area which had been examined for the Navy by the USGS, and turned down various times as an exploration zone, there being no geophysical evidence to support further efforts there.

In the July 27 ERTS picture of the area, however, USGS geologist William Fischer discovered startling characteristics which had not been apparent in the surface and aerial studies of the past. In one of the ERTS near-IR bands Fischer noticed a unique contrast between land and lake water which revealed lineations that ran along the lakes and across the plains in great

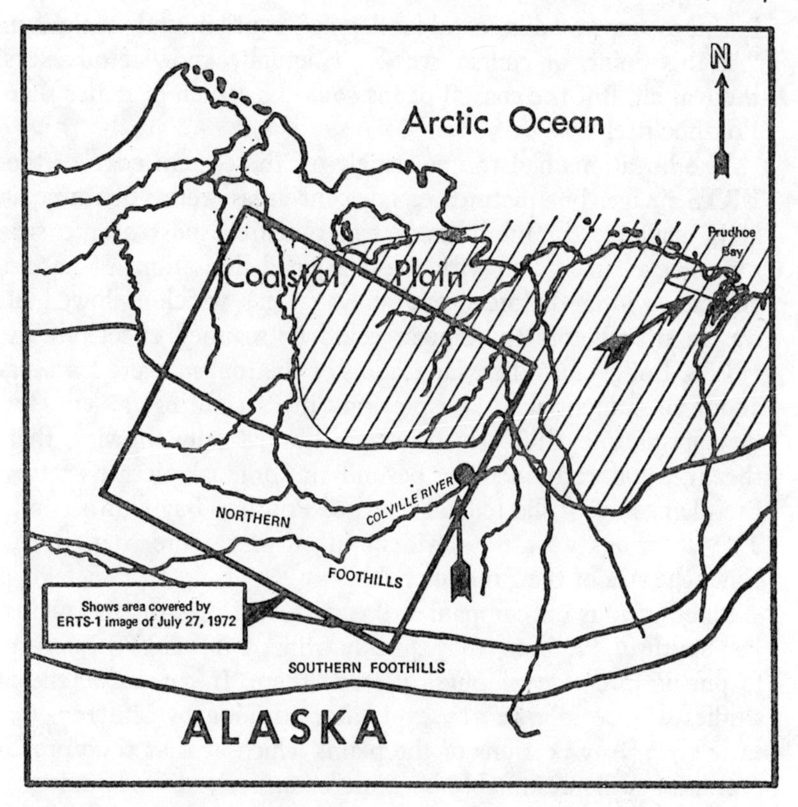

Figure 5: Areas of high oil probability (indicated by crosshatching) were discovered in a series of ERTS pictures, beginning with the outlined July 27 image. Arrows indicate the Umiat oil fields and rich Prudhoe Bay strike, which could eventually be overshadowed by the vast new areas. Courtesy U. S. Geological Survey.

numbers. He reported his discovery, and his colleague Ernest Lathram made detailed observations of the area. Together, they authored the *Oil & Gas* article, which suggested that the lineations, unrecognizable in anything less than a synoptic view, offered the potential of a super-strike. The great North Slope discoveries of the late 1960s might conceivably be duplicated or surpassed a decade later, just a few miles to the southwest.

When we discussed the situation some months after his study,

Lathram expressed a combination of caution and optimism. "At this point, of course, we don't actually know if oil exists there at all. But the coastal plains *could* be as rich or richer than Prudhoe itself."

The lineations had run enticingly off the eastern edge of the ERTS image, but pictures of adjacent areas were not immediately available; in July, clouds had obscured most of the surrounding plains. Then with the combined limitations of nature (clouds, winter darkness, and snow cover) which followed, it was months before those areas could be scanned effectively by ERTS. But even as the May readers of the *Journal* were learning about the potential of the previous July's sighting, a series of new springtime ERTS pictures was dramatically showing that the lineations extended far beyond the original picture—across the plains east of the federal lands to Prudhoe Bay and beyond. The lineations were now evident in an area more than thirty times the size of the Prudhoe fields.

Since various oil companies already held leases on the plains just south of Prudhoe, they quickly utilized the ERTS imagery to pursue intensive geophysical study there. If successful, these studies will be followed by exploratory drilling by late 1974 or early 1975. Any portions of the plains which at that time prove promising will presumably be more extensively drilled throughout the mid-1970s, with production conceivable soon thereafter. The Alaskan coastal plains (with more ERTS discoveries perhaps ahead) would seem a likely and timely successor to Prudhoe as the source of oil for the Alaskan pipeline.

The Alaskan plains area is one which was opened to worldwide view by ERTS in its very first days of scrutinizing the earth. Since that time subtle signatures in other ERTS pictures may well have attracted oil-company geologists to other parts of the Arctic region, without public knowledge, and we have recently learned that the worldwide data bank of ERTS information is definitely leading oil companies on secret explorations to other parts of the world. But whether they are in the Arctic area which has fascinated geologists Lathram and Fischer or on the opposite side of the globe, ERTS is destined to be a tool which will be used to reduce by many years the time-consuming initial search for high probability areas. On Spaceship Earth, where fossil fuel reserves are being reviewed on a year-to-year and even month-to-month basis, this time saving could offer an incredible advantage.

ERTS has offered another assist to the oil developer in Alaska. ERTS pictures of northern areas, supported by ground checks, have shown that the destruction which is known to have occurred in the oil development of the 1940s has since been largely overcome by Mother Nature, according to William Fischer. Knowledge of this ability of the terrain to recover was made available to Secretary of Interior Rogers Morton in 1973, reinforcing his argument for installing the pipeline in spite of environmental damage. This time ERTS had worked in behalf of the oilman; at other times, it would police him; in both situations, Spaceship Earth should benefit.

The availability of ERTS data should mean that the public would benefit in information about energy potential. The USGS and the Bureau of Mines, in their surveys concerning fuel and mineral probability areas, create extensive reports which are not printed until a year or more after data is fully analyzed; but from the mid-1970s on, these reports should be incorporating satellite information, providing skilled geological interpretations of remote sensing. While this more comprehensive disclosure of high probability areas will not offer a panacea for that share of the public seeking "full information" on oil companies and known reserves, it would serve as one basis for national and eventually international planning—for improving studies of Spaceship Earth's fuel supplies by Club of Rome-type groups in the future.

The Mineral Conquest of Alaska

The interest of the USGS and such geologists as Ernest Lathram in Alaska has been a continuous and fairly intensive activity since World War II. In 1964, Lathram was compiling maps of northern Alaska, pulling together the efforts of dozens of geologists who, over the past twenty years, had tramped over the tundra and the mountains in mapping expeditions. They had been productive years, in which more of Alaska had been geologically mapped than in all previous time.

A normal geological system of mapping is to divide an area into quandrangles of one degree latitude and perhaps three degrees longitude (at north Alaskan latitudes, this being roughly comparable to sixty-nine by seventy-nine miles). In the USGS, a single geologist is generally taken into the wilderness and dropped off, one geologist in each quandrangle, with instructions to map it. Hopefully, before winter falls, the geologist will have found his way back to civilization, and, after several summers of this, will have completed his quadrangle map and added it to the accumulating geological picture. "A reasonably accurate map is then created through the inductive process of assembling little pieces into a big one," describes Lathram. By 1964, enough little pieces had been contributed by individual mappers so that almost half of the immense state of Alaska was geologically mapped.

At about that same time, the U. S. Weather Service was experimenting with the Nimbus satellites, all of which followed a polar orbit comparable to that of ERTS today. Their altitudes varied, but all were higher than ERTS, and they used a wider scan angle, sweeping a much greater area than ERTS. Their business was scanning the weather, but in the process of this task they of course also brought back images of the earth beneath the clouds, and the pictures were impressive in their tremendous scope, if not in their quality.

During these same years, geologists were periodically finding themselves impressed by the eye-opening geological discoveries of Gemini and Apollo. As a result, they soon began looking for other sources of satellite pictures, and found Nimbus an interesting possibility. If Gemini, with its occasional snapshots, was useful, they reasoned, a constant scanner like Nimbus should be invaluable; why not borrow the scan images from the Weather

Service? With its frequent travel over Alaska, Nimbus seemed to offer too good an opportunity for the USGS to pass up.

Geologists were soon studying Nimbus' fuzzy, single-band pictures. By observing where snow and vegetation started and stopped, they were able to outline river channels and the fronts of some ranges, and in the process detect a few fault lines. Some of the fault lines were already known, while others were totally new discoveries. However, clouds usually covered so much of Alaska that the geologists could see only occasional short segments of each fault—not always enough to identify them firmly.

"Then on March 29, 1971, Alaska had a rare day of clear skies," Lathram recalls. "And in one impressive Nimbus picture, we could see most of the state and the western edge of Canada." Faults that had been seen as bits and pieces were suddenly joined and recognized for the first time in this one great synoptic picture. A dozen completely new faults, cutting deep lines across Alaska from northeast to southwest, were suddenly identifiable.

"Synoptic views with even relatively low image resolution are of value," wrote Lathram of the Nimbus results in *Science* magazine, March 31, 1972, adding, however, that "greater image resolution will enhance their usefulness." He was obviously looking ahead to ERTS that same year.

Lathram wasn't disappointed. In the fall of 1972, most of the first ERTS pictures were heavily cloud-covered, but on those passes when Lathram could see through holes in the overcast, he found that the four bands of ERTS were adding dramatically to all prior information. He viewed the imagery under a 30-power microscope, enlarging it to a 1:100,000 scale. In the red band, great sheets of limestone showed up distinctly against a background of shale. Ultra-mafic rocks, strong indicators of iron, magnesium, and calcium, stood out in the near-IR, and hints of chromium and nickel jutted out from the background. The multiband sensing was uncovering a variety of signatures that had gone unnoticed before. It was not only showing new fault lines, but was revealing exciting potential mineralization in both new and known fault areas.

By using satellite pictures in their mapping, USGS geologists had reversed the time-honored inductive process of making big pictures out of little ones. "By starting with a hundred-mile ERTS image and breaking it down into smaller pieces, we used *deductive* sensing," Lathram said. By looking at the big picture first, he was then able to deduce the significance of lines and patterns in the local areas. "The mapping that formerly required men to spend tens of thousands of hours, scrambling around on rugged terrain, is now being done better by working directly from a single ERTS picture!"

The mapping of Alaska was advancing by leaps and bounds, seemingly limited only by the weather. There had been enough holes in the clouds, over the first six months of ERTS, so that Lathram had accumulated 75 per cent coverage of Alaska. Only a short time before, geologists had been impressed by the fact that Alaska had been more than half-mapped in a single generation. The achievement now had suddenly been easily surpassed by ERTS in its first half-year of life, and a revolution in mapping had occurred.

X *Marks the Spot*

In one of the first ERTS pictures showing Alaska's Tanana River, Lathram noticed that a lineament which a Nimbus picture had suggested now showed up clearly on ERTS imagery. Whereas it had been a bare hint of a line on Nimbus, it now jutted out boldly. Running northwest-southeast along the river, the Tanana fault's northwesterly portion cut into the once rich Fairbanks gold area. At the opposite end, to the southeast, was the Alaska Range, with its substantial deposits of both copper and gold. In between, the new fault ran through the flat Tanana region which until now had never been suspected of mineral wealth.

The ERTS picture showed not only the Tanana fault but also several fissures which intersected it at right angles. Each of these crossings on the picture signaled one thing to a geologist—a potential trap for more of the rich gold and copper that had been mined to the north and south. Why? Because in an earlier age, molten magma rushing through one set of lateral faults had been trapped when transverse cracks broke the magma flow—leaving it to cool as gold and copper ore and await the desires of twentieth-century earthlings.

By late 1972, Lathram's phone was ringing with mining companies inquiring about the Tanana areas. Miners would now probably fly over intersections looking for subtle traces of mineralization—perhaps rust-colored gossan rock, indicating that gold, iron, and copper indeed were in bodies of ore below. Or, the miners might ground-survey an intersection. Drilling and extensive exploration at the intersections were tasks a mining company could now undertake with good cheer. And well they might. By being led to the high probability area, the miners could avoid the spending of hundreds of thousands of dollars on fruitless broad-area exploration. A mining company could now concentrate its investment on more intensive efforts to find ores in areas that, but for satellite scouting, would have gone untapped.

Such minerals as copper and zinc currently in short supply in the United States were dramatizing for us the gravity of the longer-range situation. The Club of Rome and other surveys insist that, based on man's current ability to estimate, we have supplies for only a few short decades in some fuels and metals, and somewhat greater reserves in others. Unfortunately, these can only be loose estimates.

The Roles of Fossil Fuels and Metals

How great is the untapped stock of fossil fuels and metals? If these resources were as readily identifiable on the earth's surface as the trees around a strip mine, we could look forward to inventorying them by satellite and providing a basis for future decisions affecting Spaceship Earth. Subterranean resources, however, cannot be directly measured. We cannot peer below the ground to monitor the materials themselves, since no airborne sensor outside the pages of "Superman" has yet proved proficient in the X-ray band; sensors can only provide us with guidance as to where to search further. We must look on the earth's surface for "surrogate" signals, from which we can infer that fuel or mineral wealth lies below.

Since coal and petroleum are finite, the crew on Spaceship Earth knows that it ultimately will need to find substitutes for most of the uses to which fossil fuels and metals have heretofore been directed. In the last generation, a small but increasing percentage (now approximately 5 per cent) of petroleum has been used to produce the plastics, rubbers, paints, detergents, and pharmaceuticals that comprise much of our lives. The synthetic fibers we wear, along with the airplanes, automobiles, monorail cars, medical products, and office buildings we utilize—are liberally laced with synthetics which can be derived not only from oil, but from gas and coal. Once the fossil fuel stock is depleted, the only known alternative for producing synthetics would be the (very inefficient) processing of trees or other such organic matter. Notwithstanding all the current vogue of sniping at today's "plastic society," a retreat from plastics would mean lowering the level of modern technology tremendously, eliminating products which in some cases are critical for modern man's health and environment. Although this author admits some prejudice because of professional exposure to a wide range of sophisticated plastics, he finds it a gloomy thought indeed that our grandchildren would be deprived of these modern materials, unless this generation can stop, in effect, burning them up in power plants and automobiles. A long list of today's activities, including the space program itself, would be improbable without plastics, while others could be achieved with either plastics or another equally nonrenewable resource, metal. The choice is clear: Man can use fossil fuels for a few decades of energy or a few centuries of material goods. This choice and related decisions can be better reached after a satellite sensor has supplied us with a good estimate of our underground resources.

However, regardless of how much petroleum, gas, and coal

exist, or of how we elect to use these resources, we know that substitutes will ultimately have to be found for them in all of their varied applications. Atomic energy is the most frequently mentioned fuel substitute for oil and gas in the 1980s and 1990s, but, alas, that too requires a fuel, for its fission—specifically uranium-235. Our known sources of uranium would last for several thousand years if utilized in the "breeder" reactors desired by the Atomic Energy Commission, but if we must expend U-235 on the less sophisticated reactors of today uranium could come into short supply within decades. Uranium is a product found in minerals which satellite images can help man to locate. The same general techniques which we observed directing mining companies toward assorted mineral deposits in Alaska can be used in leading man to deposits of uranium-containing minerals. Should nuclear power be expanded rapidly *without* the breeder reactor, the use of remote sensing could become quite important in this search.

Man is now not only scouring the earth for new mineral sources but is extending his search to the sun. Numerous solar-conversion experiments are in progress now, all of them in very elemental stages. So far as we know, none of these solar projects yet involve remote sensing of the earth, although satellites are being considered in other roles. One concept now under preliminary experimentation is to launch an unmanned, completely sun-synchronous satellite to collect the sunshine and relay it to earth via microwaves. Unlike solar-conversion systems located on the ground, the satellite would be constantly showered with the sun's rays—having the double advantage of being above the earth's cloud layer and being continuously stationed on the sunny side of the world.

The Earth's Birthright of Energy

Earthlings, consciously or not, have sound historical reason to extend their search for energy to the sun. Scientists generally believe that the sun gave birth to the earth some five billion years ago, promptly throwing the young planet out into the solar system to fly on its own. Ever since that time, the sun has been faithfully sending her offspring nourishment in the form of radiation. Her photons brought man immediate heat from the sun, and her light rays grew timber on earth to produce more heat. In recent times man has learned to bore into the top mile of the earth for ores, liquids and gases, which, after eons of dormancy, have been refined and combined to produce oil, coal, gas, and indirectly electrical and nuclear power.

We have, however, never yet drawn deeply from that wellspring of "natural heat" (as the geothermal scientists like to call it) which was first endowed to earth when she sprang from the womb of Mother Sun; the fire at the center of the earth, which probably reaches 9,000 degrees Fahrenheit, provides heat which we notice whenever we choose to burrow underground. Why can't man drill holes deep into the earth and capture this energy? At present, science sees no economical means of drilling deeper than half a dozen miles down (using up vast energy in the process) and collecting heat which is too broadly diffused for efficient recovery.

However, Nature herself has offered one means of assisting, by furnishing us with hot springs that flow up from the inner earth. Sometimes the source of the springs is water contained underground in volcanic formations, while in other cases it is simply rain water which has drained deeply into the earth to become heated by the earth's internal fireball, then bubbled and steamed back to surface. In either case, its vapor has been recognized as potential muscle for man to use in doing his work.

Thus far, only a few places in the world have harnessed the energy of springs into useful geothermal power. Although used in Italy as early as 1904, geothermal power has become a popular consideration in the United States only since 1960. At that time, enough steam was captured from the Geysers, eighty miles north of San Francisco, to develop some 20,000 kilowatts of power. Each year since, additional generating units were added until, by the end of 1973, approximately 396,000 kilowatts were flowing through Pacific Gas & Electric (PG&E) lines to serve some 400,000 northern Californians (a kilowatt per person being a current usage rule of thumb). By 1976, the output will have reached 908,000, enough to serve all of San Francisco.

Throughout the western United States, numerous springs and geysers are well known and would seem to offer an immense source of underground power. "Unfortunately, most of the springs are not clean enough," says Christopher Newton of the PG&E. Briny water is more often available than the dry steam which power companies require. Temperatures are often not hot enough; it takes almost 400-degree (Fahrenheit) heat to produce the pressure desired for commercial power generation. Therefore, only a handful of geothermal projects are currently planned in the United States, for the simple reason that we aren't finding enough springs that meet standards. Few power-company people are expecting geothermal sources ever to provide any great share of their needs. One leading geothermal advocate sums it up by saying, "Ten per cent of our power supply would be a lot to expect from geothermal. Our knowledge of the nature and extent of our springs is simply too inadequate to anticipate more."

The principal reason for the inadequacy of our knowledge of springs is that many of them lie unnoticed, breaking through the ground in very small springs or geysers, or, more often, existing entirely underground. The temperatures of the surrounding soil is sometimes increased, but often by only one degree Fahrenheit. Nevertheless, hopes had been held for Skylab's Multi-Spectral Scanner, with its far-IR band; the MSS, it was thought, might prove itself capable of detecting springs substantial enough to cause one or more degrees of soil heat differential, enough to be noticeable at least in isolated regions uncluttered by the heat of civilization. (This practice had already been carried out by aircraft looking for geothermal activity, with moderate success.) However, the MSS on Skylabs 1 and 2 did not achieve any such degree of performance, and the Skylab 3 sensing in far-IR has yet to be analyzed.

Geothermal Developments

Since heat variations are often so slight, it seems apparent that we cannot rely on that form of detection alone, and scientists are seeking other factors which would identify areas as geothermal. One intriguing search technique now being developed had an inauspicious beginning in July 1969, on a day Ira Bechtold and his wife were spending in the Arizona mountains. Bechtold had driven into the mountains carrying along a map, much as treasure hunters seeking the "Lost Dutchman's Mine" and other real and mythical caches had done in that part of the country for generations. But there was something rather different about Bechtold's mission; he didn't know what he was looking for.

Bechtold was trying to locate a spot on his "treasure map" (a near-IR photograph of Arizona taken five months before from Apollo 9). The photo covered a hundred-mile square, entirely blanketed by a light snow, except for one ten-mile oval that stood out from the white photograph like a hole in a bedsheet. His friends at NASA had been puzzled by the oval in the photo, and so Bechtold had taken it upon himself to check it out in person.

As it turned out, it wasn't much of a challenge. Even though things looked considerably different now, in the summer, it had taken him just twenty minutes after turning off the hard-top road to locate the oval which had, in February, rejected the snow. The reason was obvious now; the area, except for scattered vegetation, was solid rock. Like the warm sidewalks in a city, the rock had apparently melted the snow as fast as it had fallen by emitting the long-IR rays it had stored up from sunny days. It had then signaled its identity in the photograph by appearing bare and gray. The spots of vegetation, which appear red in near-IR photos, had shown up even more vividly.

After Bechtold had spent half the July afternoon hammering and chipping away at the rock formation, he and his wife retreated to the lodge in the town of Hanigan, to carry out their plan of watching the town's only television set that evening. The reason for their interest in television was not the excellent TV reception found in the isolated mountains, but the event which was to be carried on the networks that evening; they were about to observe Neal Armstrong's historic first steps on the moon.

Interestingly this day that had climaxed man's ascension to the moon had also launched a significant, if inadvertent step toward probing into the inner earth. Over the next several years Bechtold was to spend considerable time pursuing the phenomenon of circular or arced patterns that appeared in imagery after light snowfalls in the western states. He found that sometimes the arcs indicated "heat sinks," as he referred to the phenomenon he had tracked down in Arizona. Other times they connotated areas which had previously been heat sinks but which now, through geological morphology, had become sources of mineral deposits. And still other times they indicated a source of heat rising from deep within the earth—geothermal power that could perhaps be effectively harnessed. Bechtold was showing that heat from underground, which is often so slight it cannot be measured by a sensor in the sky, is nevertheless great enough so that it can melt snow faster than surrounding terrain. Then at least for a few hours a day, until the rest of the snow melts, it is uniquely identifiable.

Bechtold is now a consultant for an exploration company quietly engaged in the search for geothermal deposits. The last time I saw him in his lab in early 1974, he was studying some ERTS images of Nevada (precisely which part of Nevada he preferred not to reveal). I saw the distinct circles in the ERTS picture, and Bechtold told me that in Nevada, an area he had explored extensively on the ground, he had reason to believe that these signatures would lead him not to heat sinks—but to desert areas where valuable hot steam lay hidden beneath the sand.

Bechtold's snow enhancement work is not the only technique being developed to locate hidden hot springs. A different concept which has become increasingly popular since the first ERTS pictures is based on the realization that a unique white clay frequently surrounds geothermal activity. The clay seems to offer a particular signature noticeable in one of the ERTS near-IR bands and another signature in the red band. When these two signatures coincide for a particular area, high probability of an active geothermal zone exists. Dr. Robert Vincent of the Environmental Research Institute of Michigan has developed an automated "ratioing" system for matching the signatures in the two bands and locating those areas where the white clay anomaly exists. His system involves feeding information from each band into a computer capable of automatically matching the signatures; ideally, the computer will be able to use the ERTS input to pinpoint a vast number of potential geothermal sources quite rapidly.

Geothermal scientists have recently come to agree that hot springs may undermine portions of the western United States much more completely than ever imagined in the past. If they are correct, and if a reasonable percentage of these springs should offer clean steam, then the critical problem clearly becomes one of *finding* them. Within the next few years, probably several such techniques as those we have discussed here will be efficiently combined into multispectral sensing; hopefully, they will locate multitudinous wellsprings of steam which earthlings can use as a major source of power in the western states.

This western phenomenon wherein water or steam surges toward the surface of the earth has occupied most of the geothermal scientist's attention until now. However, a completely different possible source of hot-water energy has come to light in the 1970s, and it could conceivably overshadow all geothermal expections to date. Water which some millennia ago was compressed into clays and crystallized is believed to be contained underground within great sedimentary basins. These basins occur in many parts of the United States and the world, but the region which has attracted the greatest attention in this regard has been a basin running from near the Gulf of Mexico several hundred miles north into central Texas. Researchers foresee the possibility of using pressure to force the water out of the clay and to the surface of the earth.

"The whole earth is a seed of energy," says Ira Bechtold, and we may be finding ways in which the heat content of our spaceship itself can be made to flow through its lifelines in almost infinite supply so that mankind may never be able to expend it. Such concepts as the sedimentary geothermal theory are still highly speculative, and speed is needed to research them. An important factor in accomplishing rapid research will be an understanding of the geology of the basins, and this would seem to point to remote sensing. With basins under study which extend for hundreds of miles, we see no other way in which man can expect to obtain an overview for a guide in making exploration decisions.

The Wasted Wealth of Water Power

Another renewable resource which is not being fully utilized as a source of electricity is the water flowing on the earth's surface, easily accessible and yet constantly wasted. In most of the world, water power has been virtually untapped, and in the United States only a sixth of our electricity is generated by water—whereas we have the potential to produce at least three times that amount. Even in the Pacific Northwest states of Washington, Oregon, and Idaho, where virtually all electricity is generated by water, the gap between potential and actual power developed is tremendous. In 1972 (one of the wettest winters and springs in Northwest history), with a flow of some 152 acre-feet of water for the year, the Bonneville Power Administration allowed tens of millions of acre-feet to pass down the Columbia River without ever turning a generator (a waste that would be remembered with a wince the following year).

Yet it need not be so. The problem lies in the fact that most years billions of potential kilowatt-hours of power are flushed downstream bypassing power plants, because "hydrology is just barely a science," as one of the leaders of that field confesses. Every spring and summer, the hydrologists at dams everywhere engage in a great poker game with Nature—trying to predict the flow of water that will stream down from the mountain snow and ice each day. The stakes are high; if they keep their reservoirs too high, they risk flooding, and if too low, they fail to utilize the power potential of the water.

The greatest problem is in the difficulty of estimating the amount of water that stands in snowpacks and glaciers, waiting to trickle down the mountain watersheds. Hydrologists attempt to sample the snow throughout the watershed and measure its water content, but results have generally been less than impressive; a daily water prediction error of only 8 per cent is outstanding, and a poor day may be off by 40 per cent.

A later chapter will discuss the expected development of microwave sensors to accurately inventory the water in any given region. For the present, however, we note that Dr. Mark Meier, glaciologist for USGS, has conducted an ERTS project to measure the total water content of snow and ice in the Pacific Northwest, Canada, and Alaska. "In the state of Washington, where water is responsible for almost all the power, a satellite measuring system would be a tremendous boon," Meier said. And after experimenting with ERTS pictures for a year, Meier is enthusiastic about the ability of satellites to provide the mapping necessary from which water volume could be calculated. He found that the resolution of ERTS was adequate to study the ice and snow. However, repetitive pictures, allowing glaciologists to make calculations as to water flow, were necessary to conduct the ERTS experiment fully; and ironically, in the midst of a low rainfall year, heavy cloud cover on the days ERTS passed overhead occurred throughout the summer of 1973 and made mapping impossible!

But Meier remains optimistic about the monitoring of moisture from space, saying, "Satellites should ultimately be able to increase the water power developed in Washington by as much as 50 per cent."

As it happened, 1973 was providing an excellent example of the need for a water-monitoring program. On the seventeen days between ERTS passes, the sun shone almost continually, and the Pacific Northwest, in this year immediately following the record rainfall of 1972, was having an almost unprecedented experience—a drought, with substantial shortages of electrical power. (If the 1973 energy shortage had been contrived, then this part of it had been the cleverest strategy of all—a conspiracy with Mother Nature.) The rainfall was the lowest in recorded history, with power shortages of billions of kilowatt-hours anticipated for the Pacific Northwest. The power company understandably overestimated the Columbia's limited water flow by 20 per cent and thereby did not use the river's water to maximum efficiency; by late 1973, the Pacific Northwest, that perennial storehouse of "unlimited" dammed-up power, was one of the first areas of the country to feel the energy shortage. Responding to the pinch, Oregon led the nation in the inception of voluntary power reduction. It proved effective in reducing the power load, but it came none too soon; the same Bonneville Power Administration which in 1972 had allowed the tens of millions of acre-feet of water to be flushed down the Columbia was now buying electricity wherever it could be found, soliciting it from British Columbia, Nevada, and Montana.

By the beginning of 1974, however, the situation was reversing itself again, and the Bonneville Company was again selling power to California, as it frequently had in the past. So there you have it, from flood to drought, and then back to heavy rainfall again, all in the space of two frantic years. Man, in his poker game, was a player with a limited knowledge of Nature's deck of cards. Nature holds both the day-to-day weather and the knowledge of the changing water content of the mountains "close to the vest." We will have more to say about the emerging role of weather satellites later, but for now let us note what Meier has to say about our using remote sensing to gain a greater knowledge of that fluctuating storehouse of water, snow, and ice in the mountains. A satisfactory system remotely sensing that watershed could be enough to avoid the sort of wild gyrations in power availability that occurred in 1972–74. The 1973 period of shortage could have instead offered a surplus of electricity, in Meier's opinion, had an adequate satellite water-measuring system been in operation. Meier is continuing his study with ERTS, and is looking longingly at a future when microwave sensors, which can operate effectively in all varieties of weather, will be available to see through the clouds and monitor the moisture below.

Meier's experiment is obviously a vital one for other regions in the United States, enough so that similar moisture studies are probably to be conducted with ERTS-2 in Montana and Arizona.

Manned Flight and Space Science

Most of the experiments in monitoring the earth do indeed concern vital matters, although scientists are divided as to which projects have the greatest urgency. The average scientist, with personal characteristics much like the rest of humanity, is inclined to look on his own project as the most promising, most significant concepts ever to enter the hallowed halls of science, and he often loses awareness of other good projects. True, the scientist engaged in an ERTS investigation does feel considerable kinship with fellow scientists in space-related projects. But quite often the space scientist's camaraderie does not extend beyond the unmanned program. Many remote-sensing-oriented scientists, such as the very highly regarded USGS physicist Dr. William Campbell, look on *manned* flight as "a gigantic boondoggle, consuming NASA dollars that could be much better spent on earth observations, for the betterment of man."

A second school of thought in the scientific community, however, looks on manned and unmanned programs as interrelated and inseparable. A member of this second group is Ira Bechtold, who was occupied with thoughts on this very subject on July 25, 1971, at the unlikely hour of 4:30 A.M. Early as it was, Bechtold and a colleague had been up for an hour and were now driving through Cocoa Beach, Florida, on their way to be guests at the Apollo 15 launch. On the inland side of the road, they passed the buildings which had been occupied by outside contracting companies at the height of the Apollo boom in the 1960s. They were overgrown by weeds now, and a forlorn sign in front of one of them announced, "Space Available."

"That's an understatement," said Bechtold. There was surplus of available space, all right. A whole solar system full of it, being pretty much ignored by man. "We're failing to embark on a really great scientific odyssey into space," he was fond of saying. It seemed too early in the morning to be that prosaic now, so he pursued it differently. "Just think how earth observations have been exploding in the last two years," he said to his companion. "And all the while, the NASA program that fostered it has been withering on the vine for lack of support."

"Killing the goose that lays the golden eggs," his colleague agreed.

"I wonder," Bechtold continued, as they drove along the Florida road, "what would have happened in 1492 if the Spanish taxpayers had stopped Columbus on his way to the pier and told him to sail with one ship instead of three?" All of today's spaceships were needed, Bechtold believed. Manned and unmanned programs, he thought, were the co-progenitors of scientific advancement.

This time his colleague didn't agree. "Just think how many scientific experiments we could have going into the air this morning if Apollo's payload were carrying sensors, instead of astronauts and life-support systems."

"I don't think you would feel that way if you were a geologist," Bechtold said. He himself was sensitive to the current criticism that manned flight alone had consumed more than $25 billion in a decade, "all for a pile of rocks and moon dust." Bechtold believed that the geology of the moon had given man new concepts for studying the earth, such as a new understanding of meteor craters and their effects on the crust of the earth. In 1958, the geological community had agreed upon only 15 meteor impacts on the entire earth. Now, according to geologist Bevan French,[4] the number had soared to 57 confirmations and 107 "possibles," thanks in large measure to the use of the moon programs as a laboratory. Apart from these newly recognized impacts, the increased information on the origin of

Sudbury crater in Canada had probably been the most dramatic knowledge developed as a result of moon experience, and the understanding of this meteor had led to the rapid expansion of nearby mineral discoveries. Whether the minerals had come from the meteor itself, or from volcanic outbursts caused by the meteor, with magma spurting into faults and mineralizing, no one yet knew.

But the samples and observations the Apollo 15 astronauts would bring back would perhaps help to answer such further questions as these. In a few days, Bechtold and other geologists would be in their own homes watching TV and seeing astronauts Scott and Irwin, the earth's latest geological representatives, drilling into the moon and extracting cores containing layer upon layer of lunar history. The manned program was a geologist's delight.

The Pursuit of the Sun

The Apollo 15 countdown had progressed perfectly till its conclusion at 9:35 A.M. The ignition sequence completed, a wave of brilliant yellow flame spread out from the launch pad, and a soft murmur of response came from the crowds as the vehicle suddenly lifted, seemed to hang instantaneously in space, and then moved upward faster and faster. Bechtold found his voice increasing to a roar now, as though he were shouting the vehicle into space. His own pair of intent eyes, like a million others, was on the flaming rocket. For the moment, the rocket seemed to be speeding, not toward the moon, but toward the sun. It was a small ball of fire flying directly toward the larger one.

Bechtold was one of the tens of thousands who now lifted up cameras and followed the rocket as it seemed to chase the sun, like that earlier Apollo in his chariot. As Bechtold clicked the shutter again and again, it occurred to him that in the 1980s satellites might indeed be fulfilling the fabled Apollo role of pursuing the sun across the sky each day, with an experimental sun-synchronous satellite. And by the 1990s, satellites might conceivably be supplying a useful share of the earth's power needs from the sun. Just as, almost certainly, a satellite would be finding new fossil fuels in the crust of the earth itself. And surveying the ground for an unknown quantity of geothermal sources, and assisting in the improved harnessing of water power. Indeed, all of the world's potential systems of power would eventually depend on the synoptic information gained from space. Satellites would be undertaking as many varied efforts as possible, in a massive attempt from space to support Spaceship Earth.

The Club of Rome had given us fifty years to have new sources of life-giving power fully developed.

It wasn't a time, thought Bechtold, to hold the ships at the pier.

6. The New Ocean

By March 1973, ERTS-1 had completed more than four thousand orbits of the earth. For eight months, she had swung alternately over the north and south poles. Whereas her 115-mile-wide sensing path touched each point on the equator only at eighteen-day intervals, her coverage overlapped as she approached her intersection points near each pole, giving her the option of sensing these frigid areas many times daily. For months, however, there had been nothing but polar darkness. Then "dawn" had come, and now, in March, brief days had replaced the continuous winter blackout, and ERTS's visual and near-IR bands were able to function.

The picture it saw was an Arctic Ocean where the Soviet Union, Alaska, Canada, Greenland, and Scandinavia shared a shoreline. It was a combination of nations who were not neighbors in any east-west ocean, and there was conjecture about the future it might bring. If the ocean were to be opened, would there be dissension, or trade and cooperation?

The Economics and the Hazards

"The Arctic Ocean is the next Mediterranean," optimists were saying, and the reasons that they gave were many. The pool of oil that lay waiting to be pumped from the shores of Prudhoe Bay, gigantic as it might be, was presumed by geologists to be just the beginning. Petroleum and gas discoveries on the north slopes and in Canadian islands farther north showed signs of

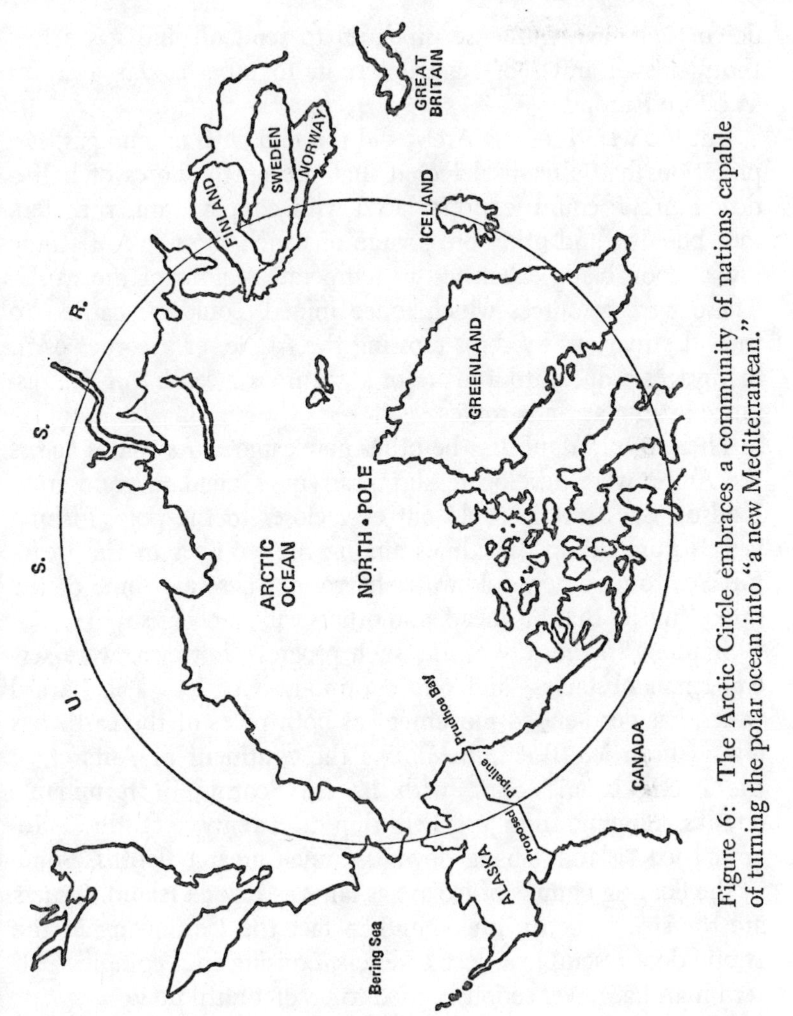

Figure 6: The Arctic Circle embraces a community of nations capable of turning the polar ocean into "a new Mediterranean."

being even greater than Prudhoe, and near these rich shorelines subsequent offshore strikes were likely to follow. "Ten or more strikes the size of Prudhoe may well be waiting," says Ernest Lathram of the USGS.

The Soviets, too, had indicated that *their* north slopes and offshore areas were wealthy in fossil fuels, and they had laid down their own immense pipelines to send oil and gas across thousands of miles of rugged terrain to users as far away as Western Europe.

But the wealth of the Arctic did not end with oil and gas. Explorations had already detected that the north shores of half a dozen Arctic countries were laced with copper, iron, tungsten, molybdenite, and other ores, some of which the Club of Rome saw as soon being exhausted in temperate regions of the world. These were resources which, once mined, could be carried to their destinations by ships crossing the Arctic, or a corner of it, in voyages reduced to a fifth (or sometimes a tenth) of the distance in regular trade routes.

There would doubtless be other new cargoes. As routes across the Arctic were developed, ships carrying Canadian grain from Hudson Bay to Asia could cut ever closer to the pole. Fishing vessels might drop their lines among the ice floes or the Beaufort Sea to reap new cold-water harvests. These are some of the opportunities that lie ahead, and others will surely arise.

Standing in the way of any such progress, however, were several small obstacles—and one enormous one. Ice. The hazard which has dead-ended movement at both poles of the earth has always been ice. But quite unlike the continent of Antarctica, the Arctic is an ocean, with its ice floating in formidable chunks, ranging from eggshell thin to twenty feet thick. Inverted ice "islands" hang down to much greater depths. Some of the floating chunks of ice are as large as Rhode Island. Others are the size of ocean liners, and in fact the ice flowing in the Arctic does resemble a fleet of liners, cruising in a complex pattern man has never seriously tried to predict until now.

The experimental icebreaker-tanker *Manhattan*, highly touted during the early excitement of the first Alaskan oil strikes, gained some idea of the might of winter ice on her maiden (and only) voyage in 1969–70.

"If she was a maiden when she left port, she didn't stay that way for long," a sailor from another vessel quipped. Reports vary as to how many times she was holed by ice before being rescued by icebreakers and allowed to slink back to her port.

Furthermore, the *Manhattan*'s attempt to blaze a Northwest Passage, from the Atlantic to Alaska, had taken her considerably south of the real Arctic hinterland, sparing her the experience of the fearsome twenty-foot ice. Her course, however, chosen in an attempt to find a lane for future oil transporting, carried her into ice a third that deep, and she scraped to a halt. The *Manhattan*, in her failure to master the ice, accentuated the severity of the problem that existed along her route and, for the moment at least, turned the oilman's interest to pipelines.

But the problem of icy seas remains one which must be conquered if we are to develop the Arctic Ocean. Indeed, even the northern shipping lanes of today are hazardous. Ice cakes go considerably farther south than the *Manhattan*'s route, floating down into Hudson Bay, or west of Alaska into the Bering Sea, where they often compete against icebreakers and cargo vessels for a piece of the ocean. They frequently win.

The Men on the Ice

The Bering Sea in March 1973 was covered by a veneer of ice chunks only a few feet in thickness. And here, hundreds of miles at sea, in the midst of ice and water, a helicopter had been frequently parked since mid-February, always with the figures of half a dozen men a short distance away, roaming over the ice.

On one typical March morning the men were working in pairs, a fundamental safety rule, and as two of them trudged carefully along the ice, one of them remarked, "If they can see us from the *Galileo*, we must look like two microcosms crawling on a frozen glass of milk." The *Galileo* was a NASA Convair 990 that would shortly be flying overhead.

The two men were approaching a large break in the ice. "A *polynya*," one of them called it, using the Russian term that had somehow become jargon among ice scientists everywhere.

These men were not concerned with the *polynya* at the moment, however, but with the ice itself. In a short while they were standing at their particular destination, bending over a selected point in the ice, and pulling upward on a small device which they had inserted in the white surface. It was an auger, and as it came upward they extracted a core of ice some four inches in diameter and almost a yard long. "Seventy-five centimeters," one of them commented on the length of the core as they lowered it into the "cold box" in which they would carry it and other samples back to the helicopter. At day's end they would return in the helicopter to the icebreaker *Staten Island*, and the core would be inspected for its salinity, its crystal structure, and its density.

The pair of men with the cold box were the last to reach the helicopter, where four other hooded figures were already standing beside a small black box. "Transit fix time," a voice came out of one of the bearded faces.

"Here it comes now," another beard added.

The six of them looked like a crew of crusty ancient mariners, but actually they were NASA scientists, and the box beside them was a mini-computer that was to print out a message, transmitted by a navy Transit satellite now passing overhead. The message would give them a fix on their location. All the data collected by each of the scientists that day pertaining to the wind, the ice, and the ocean current would be a lost cause if its location could not be accurately fixed and later used to help interpret the aerial pictures being taken of the Bering Sea that same morning.

The Transit satellite had now interrogated the computer on the ice (exactly as it regularly did to navy ships seeking navigational assistance), and had already obligingly calculated the ice cake's position. Dr. René Ramseier, group chief, looked at the paper that was coming out of the computer. The printout read, "YOUR POSITION IS 62 52 01 NORTH 176 20 21 WEST."

While the scientists watched the computer, another team of icemen was walking along on a similar cake of ice some four hundred miles away. Like the first group, they too were a weathered assortment of individuals, their faces lined from years of facing Arctic and, in some cases, Antarctic gales. One man was leaning over a small break in the ice, noting its characteristics. His reddish-brown beard was pulled together into one great icicle, pointing in toward his chest. Suddenly, he heard a sound like a fingernail scraping a blackboard which caused him to jerk his head upward. He squinted through his goggles across a horizon and saw a new break in the ice perhaps fifty yards away. "*Polynya*," he exclaimed. He, too, used the Russian word. And well he might. The redbeard and his companions were Soviet scientists.

The American and Russian icemen were involved in the same activity, gathering surface-ice data to support the remote-sensing activities concurrently under way. It was the first joint effort of the super-powers to explore near the Arctic, and it had been arranged during President Richard Nixon's historic Moscow visit only the year before. The two leading countries of the world, who recently had chased each other in the great space race, were now quietly exploring the ocean together—not in token forces, but each with a ship, a remote-sensing aircraft, and a team of scientist-explorers. (The U.S. studies would be coordinated later with information from ERTS and a U. S. Nimbus satellite.)

Previous to the Moscow meeting, in the spring of 1972, the Russians had visited the ice camp in the Arctic where Americans, Canadians, and Europeans were working on a joint study. That too had been a momentous occasion, with whisky and vodka sipped on 10,000-year-old glacier ice. But now, in 1973, the Soviets were here in force, seriously lending their expertise and partnership to a new international project—the Bering Sea Expedition, more provocatively known as BESEX.

The Airborne Laboratory

The data the Soviet and American BESEX teams on the ice were collecting, "ground truth" as it was called, was being utilized with the sensing of aircraft and satellites. The NASA

Galileo, operating out of Anchorage, Alaska, and a Soviet aircraft based in Cape Schmidt, Siberia, were sensing the areas which their respective teams on the ice surface were studying in detail. Most particularly the aircraft were equipped with microwave sensors, ideal for northern latitudes because of their ability to penetrate both darkness and cloud cover.

Sitting just aft of the pilot in the *Galileo* was Dr. William Campbell, looking out of the plane's nose at the ice conditions below. Their box pattern of flight would include the area which Ramseier and his team were ground-truthing. From middle altitudes, Campbell was able to identify the characteristics of the ice with an insight he had developed in years of camping on ice and exploring it firsthand. As his experienced eye now picked out significant breaks in the sheets of white below, he spoke into his lip mike. "Mark."

His words were carried over the intercom circuit to the after section of the plane, where technical assistant Judy Wayenburg punched a key that fed the precise time into a computer aboard the *Galileo*.

Campbell continued his transmission on the intercom circuit. "Freshly formed *polynyas* with northwest-southeast orientation. Ice has undergone recent strong convergence. *Polynyas* full of grease ice. Fifty per cent pancake ice. Remainder gray ice."

As soon as Campbell was off the intercom, Dr. Per Gloersen, sitting on the opposite side of the plane at one of the dozen microwave sensors, transmitted, "Radiometer indicates fresh *polynyas*."

While their observations were being recorded, the inertial guidance system of the *Galileo* was feeding the plane's position, altitude, and attitude into the computer. The location of the aircraft for each observation was thus plotted electronically, later to be correlated with Ramseier's ground observations, which had been even more precisely located by the Transit satellite fix. By knowing positively the character of the ice that Ramseier had encountered, and studying the airplane's microwave picture for that exact same area, reliable signatures could be developed for different types of ice; the knowledge of the ice gained in the ground truth could then be extended throughout the entire aerial picture.

Thick ice appears different in the microwave signatures than thin ice, with *polynyas* different still. The pictures that resulted as the days went on would reveal a pattern of where the ice was thick, where it was thin, and of its speed and direction of flow.

Satellite Contributions to the Arctic

The extension of ground truth, such as Ramseier was collecting, into aerial pictures had been an ongoing experiment for NASA and the U. S. Navy in several years of Arctic work. To extend the ground truth even further, the next step was up to the microwave satellite, Nimbus 5, launched in late 1972. Nimbus was, like ERTS, a polar satellite, but she was owned by and operated for U.S. weathermen. Whereas the visual and near-IR channels of ERTS could not see through the darkness of the Arctic, or through the frequent cloud cover of the Bering Sea, Nimbus microwaves could always see both areas clearly. She was a vast improvement over the single-channel visible Nimbus models with which Ernest Lathram had worked in earlier years.

The signatures being developed by the *Galileo* would hopefully assist scientists to interpret Nimbus' microwave picture of the Bering Sea and, eventually, of the entire Arctic Ocean. The thickness of ice and its direction of movement—both rather choice pieces of information for ships at sea—would be captured in the microwave images. The eventual result would be a picture rather like a topographic map; the variations in ice thickness, much like the altitude of mountains, to be indicated by false blue, yellow, orange, and brown colors. But unlike the topographic map, the ice map would be constantly changing.

Microwave resolution from a satellite is unfortunately extremely coarse. Whereas ERTS usually has a resolution of about one hundred feet (meaning, in this case, that it can detect a piece of ice as small as a hundred feet in diameter), Nimbus can distinguish nothing smaller than *fifteen miles.* The thickness of ice may, of course, vary greatly within an area of fifteen miles. But not being able to distinguish these variations within a particular fifteen-mile area, Nimbus can only establish an average thickness for the entire 15 miles.

At this point ERTS takes on considerable importance. Her ability to pick out smaller chunks of ice, if teamed with Nimbus' ability to estimate ice thickness, could combine into a reasonably accurate large-scale picture. ERTS imagery outlines every chunk of ice a hundred feet in size and Nimbus establishes an approximate thickness for it, based on a fifteen-mile average. By combining information from the two sensors, the ice can be tracked throughout the Arctic. During the daylight season, from spring until fall (when the ERTS sensors can see), physicists should be able to prepare periodic reports both on the location of ice of various thicknesses and on its direction and speed of movement.*

Eventually, when enough such observations have been made and computer-stored, mathematicians will develop formulas concerning the expected movement of ice. A mathematical model programmed into a computer, incorporating such observations as wind, current, and ice thickness, will then be able to make forecasts as to where and when the ice will move, allowing ships to determine routes through the Arctic in advance.

Vessels to Ply the Arctic

After the first Prudhoe strikes, before the pipeline was approved, various methods of transporting the oil were being explored, and several kinds of vessels were considered as tankers. (Some may yet be utilized in future Arctic strikes.) Submarines had been proven in the Arctic ever since the *Nautilus* made her under-ice crossing in 1958. A nuclear submarine-tanker was therefore proposed, but generally rejected as too expensive.

* SEASAT, a currently proposed satellite with microwave and laser gear, has the potential to further refine this process.

Simultaneously, a submarine-*tug* was designed to tow a submersible barge. It would operate on the surface, breaking through moderate ice, and would submerge, pulling its barge down with it, whenever the volume of ice required. "We can offer it in nuclear or diesel," the co-inventor, A. H. Waite, still enthusiastically maintains. And they could, he says, transport oil from Prudhoe to the East Coast of the United States at economy rates. A forty-year veteran of Arctic voyages, Waite in 1972 calculated the freight cost of his diesel submarine barge at only forty-six cents a barrel, a bargain compared to any other system.

Waite and many others continue to visualize submarines as the eventual transportation for the many cargoes other than petroleum which may be carried across the Arctic. But whether a submarine or surface vessel is the eventual Arctic solution, an understanding of the ice is essential. If ice modeling could really be accomplished, says Waite, "of course it would be useful."

"Suppose I'm in my submarine conning tower, with my radar showing nothing but ten-foot ice [which his vessel cannot penetrate] for miles ahead. Then a NASA radio report tells me where there's a three-foot-channel [which his vessel *could* handle]. Well, sure I would like it."

It was Campbell's theory that, once ice-flow modeling became available, the method of ships traversing the Arctic should become entirely different. While Campbell was hardly a seasoned ship handler, his knowledge of ice and its movement had been based on practical experience. As a scientist, he was a theoretician, but the northern reaches of Alaska and Canada, where he had lived with Eskimos as a youth, had been his first laboratory. In his twenty years of subsequent study, he had accumulated six years of time camping and working on snow and ice.

He was convinced that by utilizing models a ship could seek out areas where she could steam unimpeded through *polynyas* for as far as possible. Then, when virtually solid ice restricted free passage, the ship could hitch a ride on a cake of ice moving in the general direction of her travel, moving as part of the ice flow instead of perilously cutting her way through it. She would chart a course alternately piggybacking and moving under her own power, avoiding areas where the model indicated that ice was piling up in a high-pressure pinch.

Masters of vessels operating near the Arctic have always done a certain amount of this laborious hitchhiking, but never voluntarily. "When we are caught in the ice we move at the speed *it* chooses, which is maybe two knots," a master of thirty years said. "At that speed, we never get anywhere!" Seamen have their own ideas about operating in the Arctic, and they don't include hitchhiking.

But while physicists and sea captains have conflicting ideas on how to apply modeling, they already agree on one thing: Ice modeling would be an asset. Submariners believe that with ice models they would have a better idea of when to submerge and when to pursue thin ice on the surface. Masters of surface vessels, on the other hand, believe that with ice data at hand they could choose their routes carefully, and when necessary blitz their way through the thinner areas of ice.

As Fragile as Fierce

This blitzing concept, however, runs afoul of the environmental awareness existing today in most of the countries bordering the Arctic. Fierce as the Arctic can be when on the offensive, it is nevertheless an environment with a soft, fragile "underbelly." Campbell contends, "If the *Manhattan* had actually been carrying oil instead of water ballast when she was holed, she might have done irreparable harm to the Arctic. Her spillage not only would have destroyed the seal-hunting grounds of my old village of Tuktoyaktuk, on the north Canadian coast, but would probably also have destroyed Tuktoyaktuk itself, by melting the permafrost on which the town stands."

Campbell has been similarly concerned that a cascade of oil, spilling down the north slope of Alaska from the proposed trans-Alaskan pipeline, into the Beaufort Sea, could create a problem never experienced in temperate oceans. The algae and other bacteria that consume oil in warmer waters would not be present. The oil would survive and, according to Campbell's theory, ultimately but inevitably coat the top of the ice. He visualizes this happening in one of two ways: The oil might simply spill across the top of the ice. Or, if it were spilled in water, he would expect it to spread under the ice layer and accumulate in hollows on the rough underside of the ice. "As the ice would melt each year, from the top, and new ice formed below the oil, sandwiching it, the oil would be carried to the top of the layer, in perhaps four years."

Once the top of the ice was coated with oil, Campbell visualizes the "black ice" absorbing enough radiation in the summertime so that it would have to melt. Since the polar regions are generators of much of the world's weather, the consequences of a large polar ice melt could be catastrophic (see Chapter 11).

Enough questions about oil and ice have been raised to prompt both the Canadian and the U.S. governments to conduct tests at sea. The U. S. Coast Guard held experiments in the Arctic in the summer of 1970, intentionally spilling crude oil in the water. Observations were made of the interaction of oil, wind, water, and ice. As it turned out, Campbell's worst fears were not really borne out by the experiment; the tendency of oil to spread in the icy sea water was not as severe as he had anticipated, although the prevalence of oil accumulating on the underside of the ice, which Campbell warns about, was noted with concern.

The Coast Guard directed part of its experiment toward developing methods of cleaning up spills, and seems to have developed considerable expertise. The Coast Guard's major shortcoming apparently lies in its inability to *detect* spills. How can an oil spill in the isolated Arctic regions be detected? The long-range answer, as explained in other chapters, would seem to lie both in the microwave sensors of satellites and in UV sensing by aircraft. By the 1980s, as detailed in Chapter 13, remote sensing could conceivably be on constant alert *to detect oil slicks anywhere in the oceans of the world.*

The oil history will have other uses for satellite information quite aside from detecting slicks. The drilling for oil and the erection of offshore rigs can be a precarious operation even in temperate latitudes, but in the hostile, unfamiliar Arctic, it could be sheer pandemonium. Satellite information on the movement of ice and a satellite profile of a shifting shoreline are the kinds of information that could be as valuable to a driller as to a sea captain. The scientist can provide the remote sensing wherewithal for the user not only to explore the Arctic but also to develop technology for discovering resources and extracting them safely. Campbell, however, possessing the scientist's historical fear of the engineer with perennial plans to create the world's Eighth Wonder, looks uneasily at technology in the Arctic. "The marriage of Science and Technology is one of convenience," says Campbell (a bachelor who is perhaps suspicious of any marriage), "and frequently they are on the verge of divorce.

"Science must *lead* technology, not become her handmaiden," he adds. As an environmentalist, Campbell is concerned about the fragility of the Arctic, while as an ice authority he is excited about its opportunities and challenged by its hazards. Regarding all these things—fragileness, opportunities, and hazards—Campbell says, "Our objectives can *only* be solved through a greater understanding of the Arctic, developed by remote sensing."

Another solar authority, British explorer Commodore O. C. S. Robertson, views the Arctic rather similarly (and, like Campbell, somehow manages to attach a feminine image to it). "The Arctic is like a mistress," says Robertson. "You can do to her what she wants you to do. To try to do more is courting disaster."[1]

Remote sensing would seem to be the means of seeing what the mistress is willing to allow today, with ice modeling perhaps revealing her pleasure for the morrow.

7. Savoring the Land

It was in 1964 that national attention was first focused on student demonstrations at the University of California in Berkeley. Not far from stately Mulford Hall, which housed the School of Forestry, was Sproul Plaza, the campus area that was capturing the front pages of newspapers across the country. News cameras trained on the plaza were exposing that very first flurry of student activism—the "free speech" movement, erupting with its excited crowds of youths bearing placards and shouting four-letter words.

Within weeks of the Sproul Plaza event, student discontent waxed nationwide, ushering in what were to become years of confrontation and chaos. Shortly after that first thrust of campus dissent, however, the focal point of unrest was mercifully shifted from frank speech to other matters; and toward the end of the decade, "environment" had become one of the major objects of demonstrable wrath at Berkeley and elsewhere. By May of 1969, students and regents at the Berkeley campus had become locked in a struggle over a plot of ground which the regents considered a parking lot, but which students looked on as the "People's Park." Childish though the acts on both sides of the controversy may have seemed, the confrontation between asphalt and shrubbery was an early symbol of one of our most significant environmental issues—the management of land itself. University chancellor Roger Heyns, who was caught in the middle of the battle of the park, commented that among the various motivations involved, some persons were sincerely in the fight for the sake of "green space."[1] And on that issue the violent 1960s ended at Berkeley.

As the 1970s began, the era of demonstrations waned. Sproul Plaza quieted, and the student who was truly concerned about "green space" was more likely to be found not at the plaza but inside Mulford Hall—which, in an apparent accommodation to the 60s, now housed the School of Forestry *and Conservation*.

Among the forestry students at Berkeley in the explosive year of 1964 had been Andy Benson, an aspirant to a career as a forest ranger. In 1972, eight years and one Vietnam tour later, Benson was back at the university, this time as a seasoned staff scientist. Vietnam had ruled out his future as a ranger by inflicting a permanently disabling injury, and Benson was now dedicated to a new career oriented to the management and conservation of land in or affected by cities. It was a science which involved an exciting new technology becoming known as "remote sensing."

Inside Mulford Hall on August 1, 1972, Andy Benson was sitting in one of the forestry labs along with a roomful of other young men and women. The group had come together with a single purpose—to examine a nine-by-nine-inch color picture. Dr. Gene Thorley, the director of the Berkeley remote-sensing group, had called the staff members and graduate students together to look at their first ERTS image, a Multi-Spectral Scanner (MSS) print of the nearby Monterey peninsula.

The Berkeley group had already become convinced that the sensing of land and vegetation was the most promising application in which ERTS could be utilized. Today offered an opportunity for them to see if they had been right. "You'll like what you see," Thorley was telling the group. "As you know, we were told to expect a resolution of three hundred feet for distinguishing agricultural crops, but at first glance this print looks much better than that. I'd say it's at least as sharp as the Apollo photographs."

The nine-inch-square picture was being passed around. When it reached Andy Benson, he looked at it with a magnifying glass and studied the filtered red tones that indicated vegetation, standing out from blues and tans of the rest of the image. The picture had been taken about a week before, only three days after the ERTS launch, and the July foliage of farm crops had covered the ground around Monterey with a lush green; but in the print the near-IR bands had filtered the green foliage into a variety of red tones.

The reddish tones of the croplands interested Benson, whose own study area happened to be adjacent to the Monterey region. Benson's project was the heavily farmed San Joaquin Valley, where he hoped to assist cities in finding marginal land for their continued expansion, as opposed to the traditional urban growth practice of spilling developments out over the fast-shrinking California farmlands. Responsible persons across the nation had recently been alarmed on hearing that Massachusetts planning authorities had created a superfluous freeway which literally dead-ended into nowhere; and the same concerned public had been frustrated by nationwide examples of factory complexes covering up green belts of cropland, while nearby wastelands often remained undeveloped. Shortsighted planning was consuming agricultural land so fast, worldwide, that farm acreage would be insufficient to feed the world by the year 2000.[2]

How could this waste be recognized—not after the fact, but now, before more of it could happen? The answer, thought Benson, was to provide a satellite overview through which man could determine the best means of utilizing the lands.

If there had been a typical American concept toward land at the beginning of the Soaring Sixties, he knew, it had been "develop bigger and better." Then, in the late 1960s, "preserve the land at all costs" had been the campus response. Today, Benson hoped, the pendulum was coming to dead center. If splashing concrete across farmland and forest was wasteful, Benson knew, so also was the would-be environmentalist's concept of putting nature into a vacuum. If man were to survive on Spaceship Earth, he needed to *use* the land with its renewable resources. Land must be conserved, not preserved, just as surely as it must be used, but not used up. Benson looked at conservational planning as a means of keeping farm and forest healthy rather than idle, of sustaining the land for continued use, of savoring it, if you will, for both today and tomorrow.

He passed the ERTS image along to someone else, and looked up to notice that Dr. Robert Colwell had come into the room. Colwell was the director of all remote sensing within the university system, with six of the campuses statewide involved in sensing programs aimed at better use of California's resources. Colwell was also generally regarded as a titular dean in America's burgeoning remote-sensing community, although he looked extremely undeanish today, wearing jeans, a plaid shirt, and boots.

"I've just returned from a flight in the Cessna over part of the area in the picture you're looking at," Colwell commented. The use of aircraft photos to spot-check portions of satellite pictures was routine. An area that could be identified only as a particular shade of red on the satellite picture could be precisely identified as a vineyard or a rice field in an aerial photo. Another method of identification was to take actual treks into the field and collect ground truth in a test area firsthand. Much as in the Bering Sea ice work discussed in the last chapter, the idea at Berkeley was to define fields at close range, establish their signatures, and then use the signatures to interpret the satellite pictures. It was, someone had said, like a child in his first experience with a mirror, touching his nose and pulling his ear to identify them in his reflection. Once he knew what his ear looked like, he would recognize its "signature" evermore, and, as the theory went, so it would be in satellite sensing.

By now, Colwell was talking to the group in the lab, commenting on the surprisingly sharp resolution of the nine-inch ERTS sample picture. "Remember how much farther away ERTS is than the Apollo spacecraft was. ERTS is 570 miles off; that's like taking a picture of San Francisco from San Diego."

"Or of Washington from Boston," an eastern voice commented.

"That's a long way to be expecting a picture that's much more than a 'gee whiz' thing," one of the students said.

Another student spoke up. "You're right." Then he held up the ERTS image. "But we just got one."

Putting the Pictures to Work

It was several weeks before the distribution of ERTS data was sorted out by NASA and by EROS (Earth Resource Observational System), the Interior Department's agency responsible for distributing the pictures to user agencies. In October, when Andy Benson received the first ERTS image of his San Joaquin test area, he found it to be much like the sample picture they had seen on August 1. Under a magnifying glass, he separated the variations in color that seemed to be signatures for different kinds of vegetation. He outlined the blocks of color on the picture, discovering (as did the ice people in Chapter 6) that ERTS did indeed offer a resolution much better than three hundred feet; he could often distinguish areas as small as one hundred feet. When he finished, he glanced at his watch. It had taken him less than an hour to outline the boundaries between the various crop colors for the several hundred square miles in the test area.

Another forester and a geographer in the lab were performing the same task, independently, and Benson found that all three of them sketched substantially the same pattern. They had found seventeen variations of color in the pictures of the San Joaquin test area. He knew that all or most of the seventeen signatures would represent the various farm crops that grew in the area. But which signature represented which crop?

At this point, one of the team members went out into the San Joaquin Valley to find the answer to that question, spending a day obtaining ground truth. The ground truth revealed that one of the seventeen signatures consistently represented safflower. Safflower was almost ready for harvest in the San Joaquin Valley, and it had cooperatively appeared in golden tones in the picture. Each of the other sixteen signatures, however, were ambiguous, representing more than one crop; corn and milo maize had the same signature, as did vineyards and certain groves of trees.

However, in the months that followed, corn matured a different color from milo maize and could be distinguished from it. As ERTS made additional passes over the area, each trip differentiated between more of the crops in the valley, until finally Benson was able to identify each of them by its own unique signature.

On the same day that Benson had visually studied his first image, he had also taken a transparency of the San Joaquin area to an electronic viewer and projected it on a screen.

As he looked at the screen, he had taken hold of a lever, the "joystick."

"Kind of an electronic pencil," he had explained to a student who had accompanied him. "It's like handling the stick in a plane." The joystick had moved freely in all directions. Each time he had moved it, a trace of light had responded with the identical movement on the screen. He had electronically followed the boundary lines on his transparency, drawing a map which located safflower and the sixteen other general crop categories. In the process of this, he had been feeding his information into a computer, a process he would repeat many times in the life of ERTS-1.

As additional ERTS passes were made over the next few months, Benson returned to the machine and fed in new information, identifying more and more croplands for the electronic brain. This recognition of crops also gave him an insight to basic information about the land. Certain crops were known to require irrigation, others were known always to be nonirrigated. Thus he was able to provide the first accurate map of San Joaquin areas covered by irrigation—a contribution much appreciated by the California Water Resources Board. Similarly, by knowing what soil conditions support which crops, he was able to determine *general* soil types for the area. Again, a government agency would be informed, this time the U. S. Soil Conservation Service.

By January, the increased crop data from several ERTS passes had both augmented his information and buoyed his enthusiasm. "As time goes on, we'll learn more and more about this land. Eventually, almost any twenty-acre plot will have a long list of vital facts stacked in a data bank," Benson said. "We'll be able to point out, say, an area that could provide two harvests a year, or even three harvests." This would be the kind of land which, fifty years hence, might prove invaluable in a food-short world, land which, hopefully, would have been set aside for farming by farsighted planners of today.

Satellite Observations Fed to Computer

Later in the ERTS project, the Berkeley group began receiving an MSS tape rather than color pictures. The signatures of various crops were indicated electronically by density values in the tape before they ever reached Benson's hands. His previous chore of visually interpreting colors was thus now handled electronically. The tape was fed directly into the computer. And whereas Benson and his group had erred only 5 per cent of the time when they located safflower on the pictures visually, the electronic brain working with the tape proved to be even more accurate.

With the computer, crop information could be printed as a digital map, each crop represented by a different digit or cipher. A pageful of symbols printed by the computer could identify every crop in the valley. Figure 7 shows an actual printout of a farming area in the San Joaquin Valley where computer symbols precisely identify fields of corn, sugar beets, asparagus, and standing water. The same map for this or a larger area could be studied on an electronic screen, allowing an observer to flash as much or as little information on the screen as he wished. He might choose to study only the corn, or the sugar beets, or the asparagus areas, or only the irrigated farms, or the wooded areas, or the zones with low or high agricultural potential, providing instant analysis of the entire farming region based on a multitude of criteria.

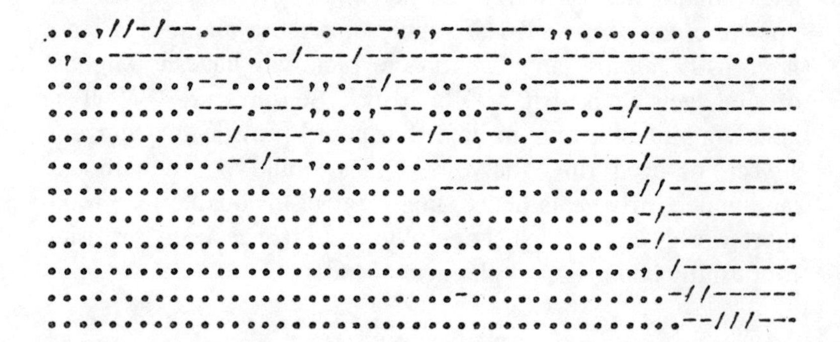

Figure 7: Satellite pictures now can be automatically processed into computer printouts. Each symbol above identifies approximately one acre of farmland in the San Joaquin Valley. Dots are corn, commas sugar beets, dashes asparagus, and slant lines water.

In the future a land planner might want to compare a dozen areas to see which had the least or most agricultural potential. With the joystick, he could outline his geographic area of interest and then ask the computer to calculate the potential value of each crop for him. He could accurately calculate how much potential food production would be lost in each farming area by opting for freeways, or factories, or houses and shopping centers.

A Satellite View of Manhattan

The long-range management of metropolitan land, as we shall continue to observe in the following pages, has become an acute problem in population centers of all sizes. Remote sensing has application to all of these, including the most metropolitan, as I came to realize one November day in 1972.

I followed Bill Harting out of the Three Bears restaurant in lower Manhattan, and we walked briskly along in the autumn air. "How about an after-lunch stroll?" Harting suggested.

I readily agreed, and we swung past City Hall and then into Nassau Street. We were able to walk right down the center of the street for a five-block area, where no motor traffic was permitted. But the street was very crowded with pedestrians.

In a few minutes, we were standing by the new twin towers of the World Trade Center. A hundred feet higher than the Empire State Building, the towers were, for this moment in 1972, the world's tallest skyscrapers*—and two more giant pinnacles in Manhattan's exploding skyline. "As people get stacked higher and higher in offices such as these, we'll continue to get more overcrowded places on the ground, like Nassau Street," Harting commented.

Finding solutions to these problems sometimes meant that planners had to look outside of Manhattan. "Remember when the New York Giants first started talking about moving to Jersey?" Harting asked.

I remembered. When the Giants had suggested building their new sports complex across the Hudson River, the New Yorkers had been furious. I asked Harting if the Jersey fans had been correspondingly happy.

"They certainly were," Harting said. "But actually, it didn't make all that much difference. For a fan here in Manhattan—let's say standing in Times Square—either stadium would be about six miles away from him.

"The point is, though, that the Giants are the home team for a metropolitan area that measures 150 miles from edge to edge. Do you know what the president of the Giants told the New Yorkers? He said, 'You're not losing a team. You're gaining a sports complex.' *He* had the regional outlook."

The 150-mile area Harting was referring to was one of the world's foremost strips of congestion. It included parts of Connecticut, New York, and New Jersey, whose combined problems Harting studied daily in his executive role at the

* They were surpassed a few months later, in early 1973, by Chicago's Sears Tower.

Tri-State Regional Planning Commission. The last few years had seen an exodus to the suburbs by major companies and tens of thousands of their employees. Thousands of decisions were made daily by municipalities, companies, and individuals regarding places to live . . . sites to build factories, power plants, or a new stadium . . . routes for rapid-transit systems, freeways, water and sewage systems; methods for handling solid waste . . . care of recreational lands . . . ways to avoid air and water pollution. There were all kinds of complex problems, virtually none of which was isolated. All were interlocking, one to another, and concerned 20 million residents, as the Giant president had realized. The tri-state commission had been formed by the three state governments to coordinate the planning efforts of no less than 600 municipalities.

They needed a synoptic view.

"We would like to think that ERTS would give us the quick overview we need," Harting was saying. We were back in his office by now, and he had pulled out a display board showing ERTS pictures of the tri-state metropolitan area. It was such an irregular-shaped area that it had taken portions of six ERTS pictures to cover its extremities.

To me it seemed inconvenient that he had to use six pictures.

"It's not ideal," admitted Harting. "But when you compare it to the airplane mosaic we used previously, it seems better. We're patching together six big pictures instead of five thousand little ones!"

Pre-Crisis Planning for Energy

One of the tasks for which Harting in 1972 was expecting to use his ERTS overview related to power plants. He and other responsible officials had for some time been anticipating an energy shortage for the tri-state area. No community in the metropolitan region wanted the brownouts, which seemed inevitable unless new power plants were established. But, alas, human nature being what it is, neither was any neighborhood anxious to accept a close-by power plant—whether of the acid-spewing coal variety or the thermal-polluting nuclear type.

With the shortages of all fossil fuels, nuclear power seems to be the most oft-discussed power solution in the United States. Nuclear reactors, however, are accused by some critics of overheating streams† and leaking radiation, Atomic Energy Commission assurances to the contrary. Local citizens with these suspicions, even though faced with an energy shortage, have often fought for plants that are isolated, downwind, and located on fair-sized bodies of water. When all these stipulations are added together, you have an immense problem, as many communities fighting the energy crisis have learned.

To find locations where authorities will agree to install plants, two kinds of regional information are needed: population forecasts to determine the power requirements of a community, and maps locating the bodies of water where plants can be established.

In the tri-state area, the first of these two tasks—population forecasting—might seem to hold little potential for ERTS, since by 1972 Harting and the planning commission had already had considerable experience in the forecasting field. However, ERTS offered a means of doing the job better. By taking the six ERTS pictures which Harting showed me in 1972, and then comparing them with a similar six-picture mosaic in the fall of 1973, forecasters for the first time would have two uniform images to compare. Population trends that would otherwise have gone unnoticed would begin to take form in the satellite overview and allow for the eventual development of models and sophisticated population forecasts.

In the second task, that of locating water, the help from ERTS was to be more dramatic. Repetitive ERTS coverage could supply a picture of lakes and streams and reveal their variations in size throughout the year. If population forecasts would indicate to the utility commission that a given area

† The trend in some areas is to use water towers for cooling, transferring the thermal pollution concern from water to air.

would need X kilowatt-hours of electricity by 1980, the utility engineers would know both the capacity required in a proposed power plant and the size of the lake needed to cool it. ERTS data could then be used to select a lake in the area that maintained sufficient water, year-round, to produce the power. From the hundreds of bodies of water in the tri-state region, a computer could sift ERTS data to select lakes which (a) were the right size, (b) were located near the community needing the power, and (c) were nevertheless apart from high-density residential areas.

I talked with Harting a year later, in November 1973, and found that both of his objectives with the ERTS pictures had progressed as expected. Areas where new developments had occurred during the year were distinguishable in comparisons of the 1972 and 1973 mosaics. Harting was preparing to have both mosaics put onto tape, so that a much more accurate electronic comparison could be made. More sophisticated population forecasting seemed clearly to be in the offing. As for locating bodies of water, Harting noted that the lakes "are one of the most distinguishable features on the ERTS images," so he now had information which could be used relative to the dual problem of energy needs and thermal pollution. The year of 1973 had been a big step forward.

Since our conversation the year before, however, a new dimension had been added to the tri-state power-plant problem. Deep-water ports, located some twelve miles offshore, were being planned to feed oil to tri-state power plants, using undersea lines. The possibilities of oil leaks required an environmental impact study that would reveal the tides and currents of potential oil drifting. The energy crisis was now upon the planning commission and others, but environmental attitudes remained influential also. An ERTS overview seemed to offer the best tool for finding solutions to both problems, and for providing a means of planning with minimal delays.

Power plants, Harting's most urgent problem, are of course just one of the interlocking requirements he is facing. The transportation arteries, the recreational areas, the industrial growth, and the residential areas all have to be considered in interrelated plans. As variables change constantly, planners have to keep their options open. Organizing the tri-state area without current information would be like assembling a jigsaw puzzle in the dark. Even when "working in the light," Harting has to spot the pieces quickly if he is to fit them together. Satellites, with the quick repetitive view, seem the best solution to his quandary.

The Wild Impact of Motorcycles

If the forces of urban living are a complex problem in themselves, they also break out onto surrounding areas with explosive impact. The Thanksgiving weekend motorcycle race from Barstow, California, to Las Vegas, a race which particularly attracts cyclists from the Los Angeles area, might be considered a case in point.

Strictly by happenstance, the 1972 race commenced just half an hour before ERTS was due to pass overhead. It was then, at nine o'clock in the morning, that some 2,500 cyclists on a mile-wide starting line revved their engines, awaited the starting flag, and then tore ahead in three waves of more than 800 cyclists each. As they sped ahead, occasionally colliding disastrously together, they created a massive trail of dust that would mark their 160-mile-long route across the desert.

ERTS arrived on the scene half an hour later and duly recorded the initial blast of dust, as did a few other official witnesses. Photographers on the ground were making a record of particular areas before and after the cyclists. Furthermore, a NASA U-2 aircraft had passed over the area a week in advance of the race for a "before" picture, and would fly over the following month for an "after" view. The Bureau of Land Management had requested the pictures to help the bureau gauge how the environment endured the experience.

The results proved to be mixed. While some terrains were virtually plowed up by the wave of churning cycles, other, tougher soils appeared to have been surprisingly immune to them. The bureau, interested in keeping future race routes away from fragile areas where eco-damage would be extensive, found that it had a valuable tool in remote sensing. ERTS and aircraft pictures in the months that followed provided repetitive views of the effects of motorcyclists and such other desert vehicles as recreational "dune buggies." A student at the University of California at Riverside, James Huning, doing some of his graduate study on the ecology of the desert, has made some interesting observations. Huning has found that the casual weekend motorcyclist, who charts his own course and crisscrosses the desert with his own trails, has had even more widespread effect on the desert than the participant in an organized race. "With repetitive remote sensing of the desert," says Huning, "we're sometimes able to discern where changes are occurring and where they are not. Remote sensing enables us to learn what damage occurs almost when it happens, rather than years later."

Significantly, the bureau in 1973 declared some areas of the desert off limits to motorcyclists and dune buggies, basing the decision in part on the sensing information. Metropolitan humanity's onslaught upon the desert has been suddenly brought into clear view, and as a result the desert is gradually being placed under protective custody before it is destroyed.

Flower in the Desert

A desert is exciting because it can sometimes be barren land one day and a flourishing cropland or a thriving metropolis the next. In certain cases, this has been almost literally true. For example, a desert-agri-urban area which attracted a number of land-management scientists long before the launching of ERTS is Phoenix, Arizona. While this was partly because the area was the subject of a very interesting Apollo 9 photograph, it was also because of the city's volatile characteristics. Phoenicians, as they like to call themselves, had experienced one of the most stunning growths in U.S. history, quadrupling their population in the 1950s alone. If the Romans couldn't build their city in a day, the Phoenicians apparently could.

So could outside developers. "Start with a clean slate and build your own city" had been a popular land-development concept of the 1960s. The retirement mecca of Sun City was planned and built just a few miles outside of Phoenix, gobbling up nearly 10,000 acres of prime cropland in the process.

It wouldn't have taken a satellite or a scientist to have pointed out in 1960 that the proposed Sun City, planned for an area where immense fields of cotton were then flourishing, would consume a wealth of farmland. But the long-range consequences of such growth were less obvious; the load that would be placed on roads, sewers, streams, and power facilities were difficult to visualize if Sun City were to be analyzed along with the other changing communities acting upon the Phoenix environment.

A comprehensive study of Phoenix, aimed at studying growth and its effects, has recently been made by Dr. John Place and the U. S. Geological Survey. The USGS used the Apollo 9 photo of Phoenix to map a 110-by-69-mile "Phoenix quadrangle." Signatures have been developed to show not only how land is being used, but what its potential may be by virtue of soil type, degree of fertility, and water availability. The USGS has already reached the point of being able not only to map cotton fields versus urban areas in the quadrangle, but also to indicate potential or valuable agricultural land which has not yet been put under cultivation.

The USGS has a computer model for the Phoenix quadrangle into which facts and figures can be fed, much as quantities are inserted into algebraic formulas. Computer models in this case are mathematical simulators which allow observers to see how the many forces acting on an environment can combine into various results. Planners can ask the computer equipped with the model how much growth will occur in an area under given conditions, and be immediately answered by a stream of figures in a computer printout. Or, if the model is asked what the effects would be of putting five thousand new acres under cultivation in a particular part of the quadrangle, the model can forecast such things as the increase in fertilizers and the polluted sediment that would flow from farms into streams, factors awaited with bated breath both by environmentalists and by the Army Corps of Engineers. These and other forecasts will constantly be revised as new ERTS information, reflecting changing conditions around Phoenix, is fed into the model every eighteen days.

In the past, information as to changes in the use of land in Phoenix, as in any area, was obtained and recorded long after the fact. "It was always a problem keeping track of what was happening," says Place. The computer, with its ability to process vast amounts of data instantly, recording it and displaying it in a variety of useful ways, was a step toward catching the data up with reality. Dazzling as the computer was, however, its output meant little if its input was nothing more than a smattering of outdated information. In such a situation, the missing link was a broad source of data.

Then along came ERTS, collecting vast quantities of information, and feeding them into the data system on earth below. With her flow of millions of bits of information, she could keep track of changes in land use, some of which were undetectable by other means, with phenomenal efficiency. If hundreds of surveyors had been continuously walking around the Phoenix quadrangle, each frantically recording mountains of facts, they could never have competed with the quantity of information that ERTS was obtaining, in the wink of an eye, every eighteen days. The missing link was suddenly in place; the satellite was providing current input, enabling the computer to fulfill its tremendous potential.

While the ability to keep current was a great leap ahead for land-use planners, computer modeling has gone even farther. It now is making it possible to take another giant step and *forecast the future growth* of an area such as Phoenix, showing how the area will grow under present trends. And, if the forecast looks gloomy, modeling will soon enable planners to develop a cheerier one. When a growth plan is proposed, the planner will ask the model questions as to how the plan would change present predictions. How much food would the area produce under the growth plan? How heavy would the transportation load be? How many people could be supported? How many jobs? Planners can select the best of all worlds—find a solution they like for 1980 and then, with the model's help, develop the plan to get there. The ERTS experiment is pointing toward a day, not far off, when a manager will be able to use a model, armed with completely current satellite data, to determine most of the advantages and disadvantages which various growth plans would impose on the resident of the 1980s, 1990s, and so on.

The Era of Instant Mapping

Satellite maps of the entire United States have been available from the Soil Conservation Service, at fees according to size, since April 1974. These basic maps are a mosaic of 595 cloud-free ERTS pictures, all taken in 1972. Land-use maps, while obviously a much more complex project, are also being developed from these and later ERTS pictures. Dr. Place foresees the eventual feeding of satellite data directly into perpetual land-use maps located in computers. The updating of maps, which has always been a titanic problem, would now be limited only by the cost and time required to reprint the maps. The current "map" at any time will be stored in a computer awaiting printing instructions. Place visualizes supplying revisions on a three-year basis in even the most remote areas of the country, and every six months in urban or fast-changing areas.

Satellites have considerable dollar appeal to persons engaged in broad-area mapping, who have long been painfully aware that a survey aircraft can easily cost $3,000 per day. In the past it has been both economically and technically unfeasible to collect the billions of aerial pictures needed to map the entire United States. For the moment, ERTS data is "free," being developed and financed by NASA, so its popularity at USGS is hardly a surprise. Presumably, however, USGS and other users will have to pay for any permanent space ventures they conceive, supporting their own satellites just as the weather people do now.

When they are forced to pay their own bills, will government agencies still find these satellites exciting? The collective answer at USGS, at least, seems to be a strong "yes." The satellite-computer team is looked on as the only economical means of achieving the immense mapping job the United States wants and needs.

This is not to say that satellites, with the resolution limitations they possess today, can provide all the detail needed for urban decisions; some aircraft pictures will certainly be required. But regional overviews, such as in the 17,500-square-mile Phoenix quadrangle, seem clearly to be the satellite's task.

Satellites will presumably lead to some much-needed standardization in mapping. Every state has at least dabbled in land-use mapping, as have such federal entities as the Department of Agriculture, the Bureau of Land Management, the Geological Survey, and the Corps of Engineers. All of these agencies, if they are to incorporate satellite data into their maps, will have to renounce their insular tendencies enough to conform to standards. The result will, hopefully, be one of more interchange of map information and fewer bureaus turning out duplicate maps for the same areas.

Forecasting Suburban Changes

Ranking high among the groups who have explored truly intriguing concepts in remote sensing and mapping is the Department of Geography at the University of California in Riverside. I became particularly aware of this on an August afternoon that began when Claude Johnson, a UCR geographer, and I drove along the Pomona Freeway, not far from his campus. We were talking about the ridge of hills that ran parallel to the freeway, and about the way this strip of concrete on which we were driving had altered things in that part of southern California. On the other side of that ridge, Johnson told me, rangeland was selling for a thousand dollars an acre—an outlandish price. The land was being bought by purchasers who would rent it out at a loss, for ten years or more if necessary, until time was ripe to develop mobile-home parks or high-rises. This all seemed natural enough—unless you happened to know, as Johnson explained, that a few years before some of the same land had been intensely farmed in irrigated orchards and dairies.

What had happened? Why had farmers abandoned their land and let it go to less productive use rather than replanting their orchards or modernizing their dairies? The classical pattern in land use, for areas around an expanding city, was for farms to be cut into smaller parcels—to switch from rangeland to orchards, for example, as the city grew out toward them. Concentric rings of land of diminishing agricultural intensity, as one traveled outward from the city, had been the pattern western sociologists had accepted for more than a century.

But here in the Walnut Valley, at least, things had happened somewhat in reverse. A study group at the Riverside campus had used aerial pictures in determining that farms had first gone against tradition by expanding *from small to large acreages,* and then later had reversed field to go urban. Dairymen had sold their acreages into less valuable rangeland, and had themselves leapfrogged farther out to the less-developed Antelope Valley. Why did this land, in valuable production one year, drop to marginal rangeland the next, only then later to skyrocket to ultra-valuable urban investments? The willingness of investors to pay high prices for land and operate it temporarily at a loss as a tax shelter might be the answer, or at least one of the factors economists would suggest. But what really interested Johnson, as a geographer, was simply that it *was* happening and, as a result of aerial monitoring, could perhaps now be usefully modeled, allowing planners to make decisions based on a knowledge of what was evolving for the area.

In nearby Perris Valley, sensing had advanced from aircraft to satellites in the monitoring of changes that were taking place, changes resulting from the California Water Plan pipeline now gushing forth in the previously dry valley. Perris Valley life style and agricultural systems were being altered by the waters in patterns not yet defined. ERTS pictures were being collected by UC Riverside, along with ground surveys, to develop an understanding of what was happening. Unlike the aircraft pictures taken of Walnut Valley, ERTS tapes could be examined electronically for changes in land use, readily developing a model that could be used to study the changes here in Perris Valley, and eventually adapted elsewhere. The real advantage of the satellite would thus lie in its ability to observe the land-use trends—not just in a valley or two, but in hundreds of experimental areas throughout the country. The pattern was markedly different in Perris Valley than in Phoenix, and studies needed to be widespread for a nation that was alarmed over its diminishing land and wanted answers to complex questions now.

Satellite-developed models, thought Johnson, could allow regional planners throughout the country to see what eventualities lay ahead for their communities. Was their area developing traditionally or "unpredictably" like Walnut Valley? If a model indicated that a farm area was doomed to development ten years hence, planners could become aware of the trend and have time to react with alternatives. Unexpected land-use changes have historically created chaos in a number of ways—starting food prices on upward spirals, creating ironic mixtures of manpower shortages and unemployment. These were the sorts of things that could be avoided if foreseen far enough in advance, and Johnson believed satellites held the key to the foresight.

Products of Urban Change

When Claude Johnson and I arrived inside the geography building at UC Riverside, we found Dr. Leonard Bowden talking with a student. "Sure," he was saying, "we can benefit from using remote sensing simply to see how land is being used right now. But as scientists we should be more intrigued with our ability to detect change and forecast the future."

The student lifted up his hand. "Wait. I believe you said something earlier about detecting *urban blight* by remote sensing?"

"Right. From our work in aircraft sensing we already know that multiband photography can send up a red flag when a neighborhood is sliding downhill. A good photo interpreter can become aware of urban blight sooner than a social researcher studying the police blotter."

"What does the interpreter see that tips him off? You surely can't see blight itself."

"He'll probably see several surrogate images of blight. Crowdedness is one. Litter is another."

Litter? I wondered. Identified from a satellite?

"Yes," Dr. Bowden answered my thoughts. "High alkalinity in soil or iron oxide in city dumps is easily detectable. Another useful surrogate is vegetation, or the lack of it. Vegetation in a neighborhood has been known to correspond precisely with income! Now, not all of these surrogates work everywhere. But formulas can be developed for particular places to tell us what is occurring socially."

The student commented, "Today a sociologist may take a census of bathtubs in a community to determine an economic level, right? So perhaps now the whole system of indicators is about to be changed. I mean, five years from now instead of checking a bathtub count in a written survey we might be looking for vegetation in a satellite picture."

"Both will be done, I'd say," Bowden answered. "However, the satellite information can be handled electronically and automatically. That's quite an advantage.

"But," Bowden continued, "I started to tell you that we scientists should be more interested in seeing the future than the present."

"You're saying we can make models that will actually forecast blight?" the student asked.

"Frequently."

"How?"

"Oh, by the position of land and buildings and people. How they relate to transportation arteries, for one thing. You can tell from remote sensing which will eventually be the 'right side' and 'wrong side of the tracks.' Or wrong side of the freeway, more likely." Sociologists already agree on some of the factors which cause a poverty pocket to develop in a city, Bowden explained, continuing, "but the remote-sensing overview actually allows them to see the causes *as* they develop, long before a ground-oriented planner would understand what factors were involved. You can look at successive ERTS pictures and see all kinds of developments on collision courses. And you have time to do something about them."

The student thoughtfully stroked his beard. Three months before, he had called the space program a wild, jingoistic extravaganza. Now he was hearing about satellites capable of mapping the entire United States, revealing how land was being used, and even providing an opportunity for enlightened social planning. Furthermore, the potential existed for observing trends, projecting the future, alerting planners as to what lay ahead, and perhaps even responding to poverty problems before they occurred. "A lot of people are apprehensive about computers taking over the world," the student said, only half smiling. "You know what I think? Satellites could be man's best chance of actually heading Big Brother off at the pass—reversing the fable so that *we* control *him*."

8. Search for Food

There is a great deal of talk today about the ability of the world to feed the population which is exponentially erupting over its surface.

Such groups as the Food and Agricultural Organization of the United Nations (FAO) have for years been gathering data about both our needs and our sources of food. Most of the thoughtful people who read FAO reports pronounce the long-range future as dark indeed. The world is adequately feeding only a fourth of its population now, they point out. And, they note, as people continue to double, quadruple, octuple, etc., both the well-fed and the undernourished will find their menus drastically curtailed, man being alarmingly less efficient in producing nutrients than in reproducing himself.

There are other aware persons who read the same FAO reports and cheerily foresee a world that will be unflaggingly supporting 36 billion persons, twelve times our present population, a century from now. We like the undaunted philosophy of these upbeat folk, but have trouble pinning them down on any method of increasing food at a rate comparable to the dizzying population growth. Rather than specifying food sources, these optimists talk obscurely about breakthroughs in any of several ways. They offhandedly mention such exotic concepts as hydroculture and test-tube foods, but focus their greatest expectations on (1) harvesting the "inexhaustible sea," (2) continuing to make land crops more and more productive, and (3) cultivating virgin lands.

Regardless of their degree of optimism or pessimism, most

students of the food problem today approach it on a global basis. They are looking at the problems of both the poorly fed and the better fed societies as one master dilemma, requiring a common solution. Whether one's concern is the starvation in Bangladesh or the weekly food bill in Pittsburgh, the hopes for improvement would seem to lie in the same agricultural science and technology.

Countdown for Abundance

A group of businessmen involved in one element of this technology met several years ago in Mannheim, Germany. One of America's and the world's largest manufacturers of farm equipment, the John Deere organization, had invited its key distributors from throughout half the world to Mannheim. Although the meeting was in Germany (the manufacturing center for some of the company's latest equipment), the distributors in attendance represented less-developed countries of the world, in most cases Latin America.

The year was 1967, when the U.S. space program was at its peak in popularity, and the theme in a talk by senior vice president George French had both a current and an urgent ring to it: "Countdown for Abundance." The executive was pointing to the urgency of the world food problem and the roles that both John Deere and its distributors should agree to play in it. "Here in this valley of the Rhine," he said, referring to a place where treaties had been signed and sealed for centuries, "a compact for world survival is needed."

French was familiar with FAO figures, along with a wealth of other food data and nutritional theories. He mentioned the oft-discussed food panacea, the ocean, but then, in a few brief words, discounted it as an immediate source of nutrition (however, see Chapter 10).

He went on to the second popular concept, the improvement of crops, pointing out that this was a responsibility industry must partially shoulder. His own company, cooperating with manufacturers of seed and chemicals, must pursue continued development of new farming practices. There were, he pointed out, revolutionary concepts for improving crop yields being researched at Deere, some of which were to be released to farmers that very year, with others not likely to emerge from the laboratory until the 1980s.

Then he shifted to the third popular concept. "Most of the world's unused cultivable land lies in countries represented in this room."

The vice president's audience seemed to murmur agreement. The land, they knew well enough, was needed in the fight against hunger that was occurring in each of their countries, only a few of which were net exporters of food.

These distributors were influential men in their respective homelands and might be able to influence land development, French suggested, but it seemed to me that he selected his words gingerly. On the one hand, he knew that the concept he discussed was popular with the group—good business, in fact; the distributors sold not only the equipment that could farm new land but also the bulldozers that must first clear it. Furthermore, *reforma agraria* lay ahead in several of the Latin American countries present, and while in actual practice this usually only meant changing the ownership of existing farms, agrarian reform *could* mean the opening up of a jungle. This would create a sudden need for the skidders and log loaders and plows the Deere dealers all wanted to sell.

But on the other hand, every man in the room knew that the easy land had long ago been opened. To find land in 1967 that was not prohibitively mountainous or rocky, where the cost of clearing out the jungle was not ten times the value of the land itself, where weather was acceptable, where satisfactory water and soil and drainage were combined—this would be an extraordinary goal to realize. All of these men, both the American hosts and the international guests, were from a practical mold. All had walked in the forests or the jungles of the countries in question; there were no theoreticians present who thought in terms of magically clearing a billion acres of untamed backland. Not surprisingly, the meeting disbanded with no false promises or expectations.

Finding the World's Hidden Croplands

In the years since that gathering, the Deere organization and many others have followed through on the commitments made that day. Industry has introduced all variety of innovation for putting soil and seed together as more efficient generators that come up with increased crop yields. Fine and dandy. But as for the really tough challenge, which George French had handed to his guests, not much has been done. Precious little land has been developed among the countries which in 1967 had been represented at the table by the Rhine.

Along with political problems and human inertia, the challenge of even *finding* land within wilderness areas that could economically be turned into farmland remains the first stumbling block confronting world food progress.

Why has it proved so difficult to locate tillable land? Is new land virtually nonexistent? Or simply hard to identify?

Regardless of data provided by FAO, the Club of Rome, or anyone else, *no one really knows.* The extent to which land is available will remain a secret until satellite inventories of the globe have been taken and interpreted. The identifying of terrain and its potential is a problem which has not been completely solved anywhere, not even in such North American market baskets as South Dakota and Indiana. Recognizing the problem, soils scientists in these two states have, in recent years, sought new ways of determining the soil potential of their land, including remote sensing.

The time-honored way of studying soils, of course, has been the examining of soil samples in laboratories to determine their mineral content. In 1936, however, the U. S. Soil Conservation Service was created in an effort to keep the United States from eroding down the Mississippi River as a result of floods and wind storms that not only ravaged the countryside but carried off its topsoil. SCS was assigned a task of soils study aimed at flood control and erosion, and aerial surveys quickly became important. These projects provided what to this very day are, for many parts of South Dakota, the only aerial maps, painstakingly acquired in the 1930s by aircraft photographers shooting strips no more than three miles wide. These mosaics, which of course were simple black and white, single band, provided an idea of general characteristics of the terrain, but hardly enough to tell anyone much about what crops could be grown where.

Among the rural residents of South Dakota today are several Indian tribes, all descendants of the great Sioux nations, whose life style falls considerably short of that attributed to the resplendent Sioux legend. As a result, Indians have held camp-ins in the Bureau of Indian Affairs Washington headquarters in 1972 and in Wounded Knee, South Dakota, in 1973. The problems seem to be complexly embroiled in the political stances of the bureau and the Sioux tribal council. But politics aside, one great leap forward could be made if the Indians could develop a healthy agricultural economy.

In order to determine what farming opportunities were available, the bureau would first need an accurate, current map of the soils and terrain of the Indian reservations. There might be valuable cropland that had never been farmed, scattered throughout the reservation. But how to find it? A century ago there had been plenty of good idle land, flat as a table. But now the desirable land was scattered throughout the reservation acreages and was many times as hard to identify. Locating it by trial and error would not be easy, and farming in South Dakota was difficult enough without experimenting in the field—trying to drive a tractor on grades which turn out to be dangerously steep, pulling a plow through land where bedrock lies hidden just under the surface, or becoming mired up in unexpected wet gullies.

Mapping the terrain and pointing out these "farming pitfalls" were important if Indians were ever to expand their farm acreages.

Mudholes—Color Them Blue

Well in advance of ERTS, Dr. Charles Frazee, soils scientist at the State University of South Dakota, decided to develop the university's remote-sensing expertise in studying soils on Pine Ridge Indian Reservation. NASA agreed to help.

By using a camera system that collected data in the same wave bands as ERTS, the university team began photographing selected strips of land from an aircraft and "ground-truthing" them by jeep. By observing dense spots in one or more spectral bands, and then driving to the points in a jeep, signatures for various soils were established, some of them sandy loam suitable for farming, others soil types incapable of supporting crops.

Other distinctive signs in the pictures led the jeep to the pitfalls—to the gullies and mudholes and terrain that was too steep to be cultivated, or to shale and bedrock areas through which no plow could pass. As ground-truthing discovered more and more such areas, an additional signature for each problem situation was determined.

Once the signatures were established, the conditions could be arbitrarily assigned false colors for automated viewing; blue was assigned to wet areas, yellow to slight erosion hazards, green to worse hazards, etc., all of which showed up vividly on an electronic viewer and identified farmland as either tillable or not.

After the university team had developed its technique with aircraft, Frazee began to utilize ERTS data in 1972. "How did it work?" he was asked.

"Well enough to distinguish the general soil type for an area the size of a typical South Dakota farm," said Frazee. "Well enough to map the terrain and see how much irrigable land is on a given part of the Indian reservation."

acres of grapes in an area, and then by simple multiplication to project a reasonably accurate harvest forecast.

But to guess where the grapes might be sent *after* harvest, what the farmer might or might not do with them, is enough to grow gray hairs on the head of any statistician. In the case of the Thompsons, a Fresno vineyard owner has the two principal options. He can sell his grapes either as raisins or as fortifiers for dessert wines; and, being a businessman intent on survival, he has been known to watch the prices of each market carefully, so that at the eleventh hour he could guess the best way to jump. If the income from making raisins looked promising, he might quite naturally lay out more grapes to dry than usual. Instead of loading them into a truck for the winery, he could quickly hammer together a few more drying trays to take into his vineyard, fill with grapes, and lay out between the rows where the fruit would wither into raisins.

However, no single vineyard owner could be clairvoyant enough to guess how many grapes the several thousand *other* free-thinking growers had laid out in the San Joaquin Valley that same year. Some years, the growers would respond to high raisin prices, glut that market, and run short of grapes to sell the winery. In other seasons, precisely the reverse would occur.

"It seemed to be a situation where sometimes everyone lost," remembers Ward Henderson, a USDA Reporting Service official. "Almost inevitably one market would be oversupplied, the other caught short." Farmers would then have either surplus grapes or surplus raisins to sell for some low-paying commercial use. The consumer, in turn, might find raisins or wine absent from his store shelves, or at least higher-priced. The need for some accurate control was apparent.

Knowing the limitations of written agricultural surveys, the California Wine Advisory Board commissioned the USDA Reporting Service to develop a new, more direct kind of survey, some way to maintain an up-to-date status report. The plan ultimately chosen to achieve this was to fly a photo reconnaissance plane over the vineyards and *count the raisins* (or rather, count the trays) as they lay soaking up the sun in the vineyards.

Reconnaissance aircraft at 17,000 feet began flying over the valley, week after week, photographing all the Thompson vineyards. The film was immediately processed and the totals publicly announced the following day.

Locally, it had the excitement and suspense of a Berlin airlift, with everyone in the valley part of it. Rural California seemed to have "gone Hollywood." Aerial crop pictures had been turned into a weekly drama.

It was an incredibly direct kind of census. "Like counting the pickles in a barrel, but it worked wonders," said a USDA surveyor. The system was used for most of the 1960s until the expanded wine industry demanded more grapes than could be produced. During one of the years in its heyday, *an estimated $5,000,000 in revenue was saved in exchange for a survey cost of* $50,000[1] according to one close observer.

The fact that aircraft would be considered necessary for inventorying just a single crop in one valley, with only a few thousand producers, perhaps indicates some hint of the lack of predictability in farming.

If we expand that situation into a national picture, we find millions of farmers growing hundreds of products. The problem of getting a proper mix of foodstuffs on the market, to feed the entire population at prices satisfactory for farmers and consumers, rich and poor, presents a staggering challenge which has created headaches for more than one administration in Washington. The need to keep track of the food supply—how much is planted and how much will be harvested—would obviously be as useful on a national scale as was counting raisins in Fresno. The chore of counting acreage of any and all crops is an assignment which satellites could readily handle, and probably will, thus inventorying the crops farmers elect to plant for the year. However, human decisions are not the only factor which will determine the amount of food being produced, yield variations also being dramatically affected by the fickle factors of weather, blights, and insects. Is there any way that satellites can monitor these factors?

Weather and the Green Line

In early November 1972 a visitor drove up to the Space Science Laboratory in Berkeley, an installation built by NASA and administered by the university. The lab is located high on a hill, with the bay below in the distance, and the campus itself spread out at the foot of the hill. Mulford Hall, serving Dr. Colwell and the university remote-sensing staff, is in the middle of the panorama.

As he drove up the road, the visitor looked at the grass covering the embankments on either side of the spiraling road. Whereas it had been brown on his last visit, it was green now, and that, really, was the reason he was here. He was a cattle-company executive, and green grass was the fuel that ran his business.

At the top of the hill, he parked and went into the lab. There he found Dr. David Carneggie, whose career specialty was doing just what the visitor himself had been doing on the drive up—looking at the color of grass and deciding what it would mean to cattle. Carneggie, in fact, was in the midst of that task when the cattleman entered his office, except that he was looking, not at the grass outside his window, but at an ERTS transparency laid out on a light table.

"These look great," Carneggie commented, after he had said hello. "The two ridges just east of this lab showed up on this October 6 image as having begun to turn green. So did the northern California coast from here up to Eureka.

"And now, look at this." He moved aside from the table and the ERTS pictures, so that the visitor could see them. There were two images, labeled October 6 and October 24.

Green grass appearing red as it did, the October 6 picture already showed a pinkish cast from the bay area north up the coast, where Carneggie had said growth was beginning.

Then Carneggie pointed to the later picture. "Remember the storm that rained out the third game of the World Series in Oakland? That rain wasn't all bad; you're looking at some of its benefits right now." The October 24 picture, taken a week after the game, showed that the rain had done two things. It had fallen over half of the state and turned most of the northern California image pink. And in the area along the coast, which had been rained on previously, it had increased the growth of the grass so that it now appeared a deep red. All of northern California was thus tinted either pink or red, in sharp contrast to southern California, where still-parched brown grass showed up a dull gray in the ERTS picture.

"This green grass up north is the earliest start we've had anywhere in the state for years," Carneggie said. "But there's a long season ahead, and it could turn sour. The name of the game is to rain early, rain often, and rain late. So far, we've only had the 'rain early' part."

He pulled out three pieces of paper, each containing a graph. "I've done some calculations for the grass a few miles east of here in Contra Costa County. I've based it on the World Series rain you see in the October 24 picture."

He pointed at the first page. He had made calculations on the basis that the rain would continue pouring down till late spring, delaying the crop in its maturing. The grass would not tassel and reach what he called "maximum photogenic stage" in the ERTS pictures until June 1. This long season would mean a colossal grass year for Contra Costa County—about 8,000 pounds per acre, he calculated.

"Now, here's a less optimistic look." He held up a second graph, which projected a shorter rain, with photogenic stage in the ERTS pictures arriving April 15, and the grass crop for most of Contra Costa County reaching only 5,000 pounds. "That's an average crop."

The third paper was even less encouraging. It had the rain halting in mid-winter, and limited the harvest to 2,000 pounds.

"This information will be published in your cattle-industry newsletter," Carneggie explained to his visitor. "If it shows

Contra Costa cattlemen that they're facing a sickly 2,000-pound harvest, they'll read on to see how the picture looks farther north, perhaps in Idaho. They could probably haul their cattle north and rent rangeland there. On the other hand, if ERTS reports that things are no better up there than here, our Contra Costa ranchers will plan on buying supplemental feed, or else selling off some of the calves early.

"If we have a 5,000-pound prediction, though, the ranchers can stand pat. That kind of average yield should take care of all their grazing right here in Contra Costa County.

"And if they *really* get drenched this winter and hit 8,000 pounds, they can turn around and be landlords and rent land for cattle hauled in from a dry region. The Southwest perhaps."

Everyone wanted to keep cattle as long as possible before selling to the feeder. Carneggie's visitor was no exception. He represented a large national cattle company which, as the saying went, "moved truckloads of cattle like checkers on a board." It looked to him as though Carneggie's ERTS picture and his formula would make it possible for the cattle industry to plan ahead. And whereas a certain amount of rangeland traditionally went undergrazed, satellites seemed to be capable of helping him put more of the grass and cattle together for maximum efficiency nationwide.

"You have the world's only picture of a whole ranging community," he commented to Carneggie as he left. The ERTS picture was the closest thing to a crystal ball he had seen yet, and he expected to keep an eye on it.

Rangeland and Meat Prices

He wasn't alone. Another organization which had begun paying attention to satellite pictures was Cattle Fax, a national company paid a fee by cattlemen for information on where to sell their calves for the best price.

Boom-and-bust is the usual pattern in the cattle business, with weather one of the governing factors. Tom Beall of the Cattle

Fax Denver office blames the droughts in the Southwest in 1970, 1971, and 1972 for some of the subsequent market gyrations. "The states that have been having droughts customarily range 30 to 40 per cent of all the cattle in the country," Beall commented to me in August 1972. "Last year, when range was scarce, a lot of the cattlemen in those states shipped cows off to the slaughterhouse. Those were cows that would have had calves for the market in 1973 and 1974. If beef prices are high in those years, we can be sure that the calf scarcity, brought on by range shortage, will be part of the reason."

Grass gives out all at once in some parts of the country, he explained, "by finding the unused ranges here and there, we should be able to keep from selling all our calves to feeders at the same time. The market will have a chance to level out. Obviously, the more grass we can find and use, the cheaper we can grow beef."

Few applications lend themselves as well to satellite sensing as the monitoring of grassland. Spread over vast areas as it is, in constantly shifting green and brown patterns of varying nutritional value, rangeland is an asset which will pay dividends if man is astute enough to collect them. But every acre allowed to go ungrazed is a source of meat wantonly wasted.

Because rangelands are so extensive, comprising most of the land in several western states, their contribution to the nation's larder is substantial. Today's city dweller finds this extremely hard to believe; the hundreds of miles of brown scraggly grass he may have driven past on a summer's vacation impressed him as being little more than a wasteland.

And, if we scrutinize the rangeland situation on a yield-per-acre basis, the vacationer's analysis is accurate. Depending on the area, a single calf requires from about ten to a hundred acres of grazing. Or, putting it another way, one acre can contribute no more than twenty pounds of dressed beef a year to the economy—and perhaps as little as two pounds. This is certainly making minimal use of land, albeit land which would otherwise go unused.

However, in irrigated parts of the West, there are green oases

1-a: Startling knowledge of numerous faults cutting across the earth's surface resulted when satellite photos were pieced together, beginning with Gemini 4 in 1965. Revelations have since followed in studies of earthquake origins, and of oil and gas deposits. (Photo courtesy U. S. Geological Survey.)

1-b: The eyes of the world remained on the moon throughout the 1960s, as depicted at right by NASA leaders Dr. Christopher Kraft, George Low, and Dr. Robert Gilruth on March 6, 1969. But the most meaningful items acquired by Apollo astronauts that same week were pictures not of the moon, but the earth. (Photo courtesy NASA.)

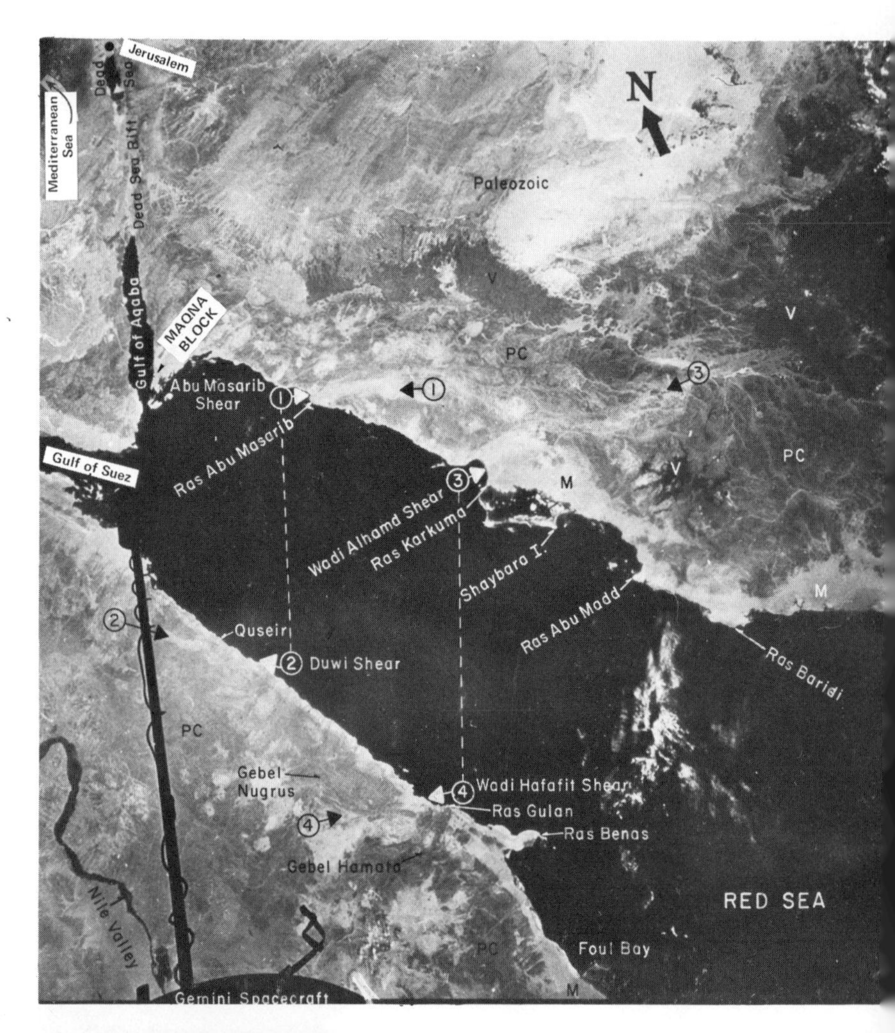

2-a: Faultlines "1" and "3" on northern shore of the Red Sea were discovered to match "2" and "4" on opposing shore. Resulting theories relate to continental drift and to 1969 natural gas discovery at Maqna Block, indicated on north shore. Similar techniques are now being applied to energy search on a global scale, creating exciting advances in a search for new oil. (Photo courtesy American Association of Petroleum Geologists and Dr. Abdel-Gawad.)

2-b: The concepts of global satellite exploration of all types have led to strong international interest. NASA's A. J. Calio, in background, confers with three Soviet scientists at Johnson Space Center in 1971. (Photo courtesy NASA.)

3-a: Rich copper area in Arizona was discovered without help from satellites, but geologist Ira Bechtold says that Apollo picture supports the high expectations of the area. (Photo courtesy of William L. Simmons.)

3-b: Pictures such as this 1973 Skylab view of Hoover Dam area are being examined discreetly by miners and geo-thermal developers, and openly by seekers of irrigation water—who see hints of underground springs in area south of Lake Mead, lower right.

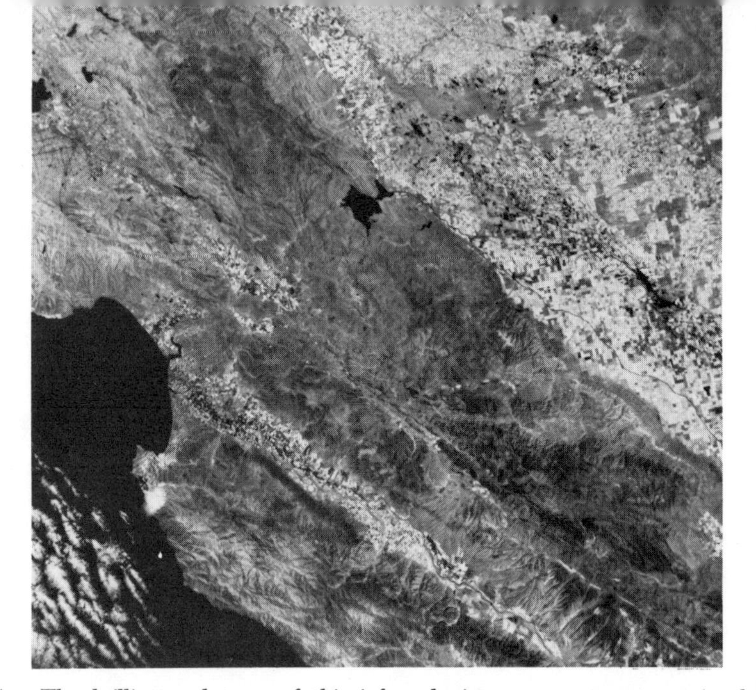

4-a: The brilliant red tones of this infra-red picture are not apparent in the black-and-white version above. However, the trained geological eye can detect numerous faults running throughout this earthquake area, the Monterey Peninsula of California. 4-b: Below, a chart for the identical area shows solid lines for well-known faults. Equally numerous broken lines indicate those faults discovered from the satellite picture. (ERTS picture and chart courtesy NASA.)

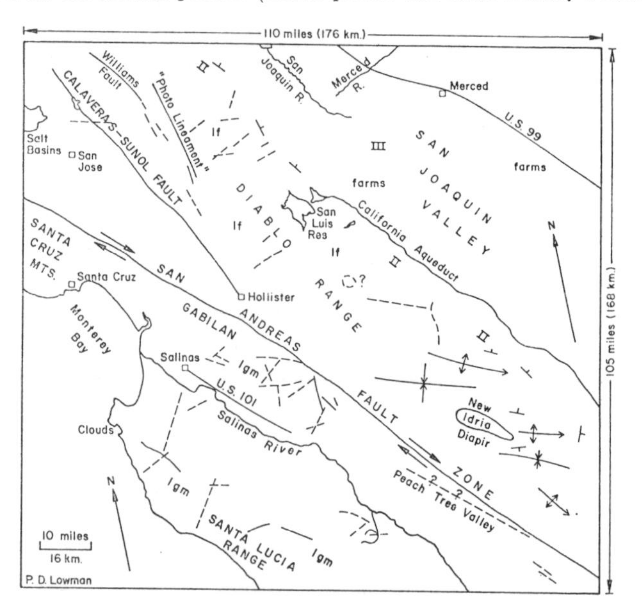

(Photo courtesy Robert Asher.)

5-a: Ice physicist Dr. William Campbell, above, has spent a lifetime exploring such areas as the Alaskan glacial region shown in satellite picture 5-b at right. Campbell contends that such pictures indicate flow and movement of glaciers and are revolutionizing ideas of glacial effect on climate and environment.

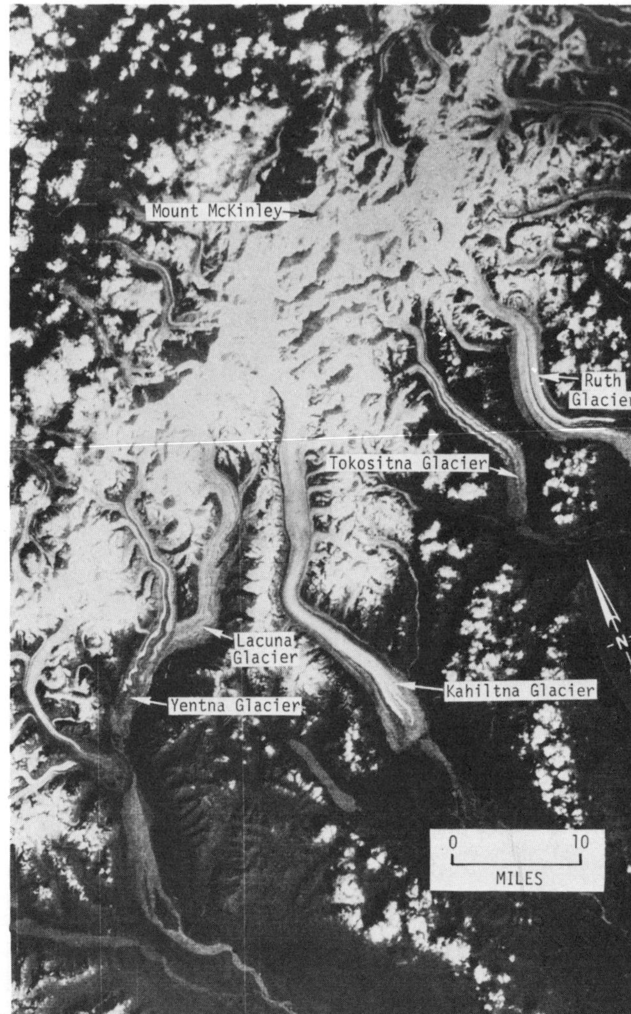

(Modified ERTS picture courtesy U. S. Geological Survey.)

6-a: Behavior of glaciers is further monitored by Earth Resources Technological Satellite (ERTS) pictures showing glacial sediment plumes extending more than thirty miles to sea. (Courtesy U. S. Geological Survey.)

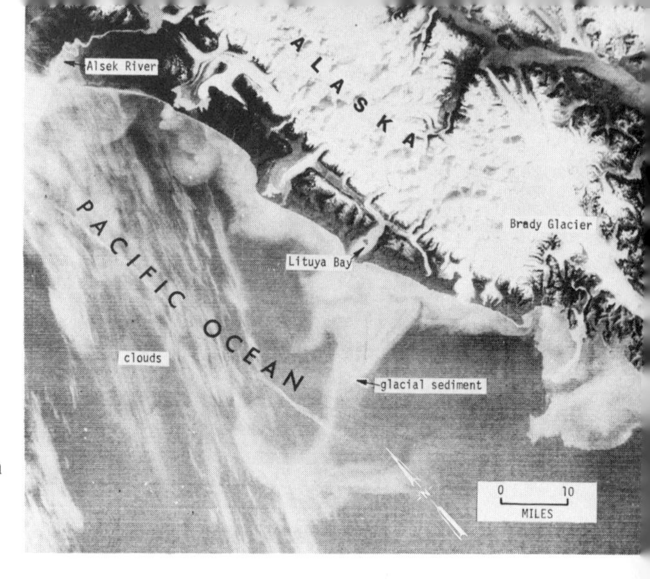

6-b: Below, a completely different application of sediment detection reveals acid wake of controversial dumping by garbage ship, south of Long Island. (ERTS picture courtesy of NASA.)

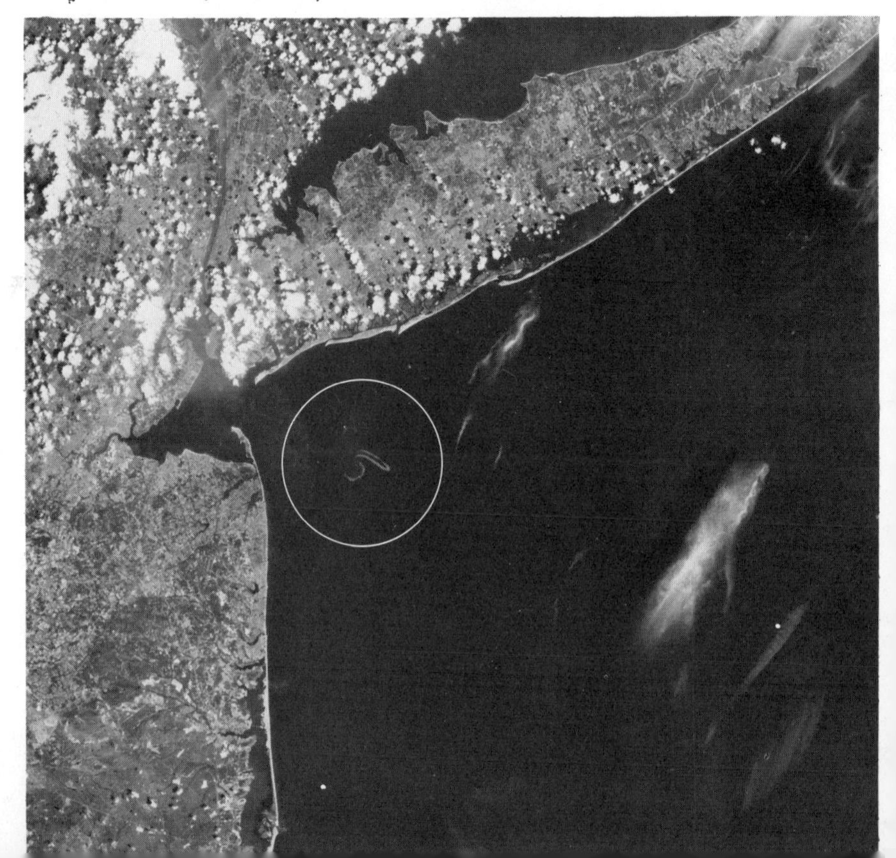

: As Secretary of Interior, Stewart Udall (left)) was early supporter of te sensing programs, as was Dr. William Pecora, director of Interior Depart- 's Geological Survey. Remote sensing in 1974 is being used to monitor eco- l damages following motorcycle races across Mojave Desert. Limitation of open to cyclists is resulting. (Photos courtesy U. S. Department of Interior.)

8-a, b: Since 1972, the Earth Resources Technological Satellite (ERTS) has been scanning the globe somewhat like a color TV camera, with each color band capable of detecting different characteristics of the earth. Even in small black-and-white reprints, as in views of Los Angeles at right, differences are visible. Agricultural growth appears dark in blue-green band (above), and light in lower picture, taken in infra-red (IR) band. Concrete-lined flood control channels running through metropolitan area are visible only in above view; city streets, conversely, are prominent only in lower.

Striking electronic displays can vary all four ERTS color bands in search of earth resources, as demonstrated by Dr. Robert Colwell of the University of California at Berkeley.

(Satellite pictures courtesy of NASA.

(Photo courtesy of University of California

of crops flourishing where the dry grasses once grew. The difference lies in the addition of that vital ingredient, water. As satellite sensing progresses and agricultural technology advances, as fertile soils are located through remote sensing and as new areas are provided with water, croplands will begin to blossom in place of some of today's stretches of grassland.

New croplands may not yet be urgently needed in the United States, but they soon will be. Some students of nutrition believe that a century from now, with a world population potentially increased by twelve times, meat will have long since become too much of a luxury to be allowed the extensive share of the earth's surface it enjoys today. Man, they like to warn, will have to cease his role as a carnivorous animal early in the twenty-first century.

In the United States, however, the situation is not so extreme as elsewhere. The American meat industry decades ago began leading the way to producing more meat with less acreage. After roaming the ranges for the first year of their lives, American beef cattle are virtually all herded into feedlots where nutrients are abundantly poured in front of them: eating and weight gaining become their way of life. Hogs and lambs are similarly handled. Under the American system, a single acre of ground can manufacture substantially more meat than under the time-honored custom of keeping cattle on the range. Instead of producing twenty pounds of meat from an acre of rangeland, an acre planted to corn can be harvested, carried to the feedlot, and converted into four hundred pounds of dressed beef.

Corn, this primary concentrate used to fatten animals, has grown steadily in importance in this country, for an increasing number of uses, ever since the Indian first introduced it to the paleface. The meager crops of those colonial days had been mostly consumed directly by the Indians, as corn meal, occasionally dismaying the red man by rotting during winter storage. It was the white man, in all his wisdom, who led the way in using fermented corn for what we now call bourbon. This was just the beginning of varied applications to which corn was eventually to be dedicated. Today it is funneled into more than twenty major foodstuffs, ranging from cooking oils to syrup, and fully as many nonfood products, ranging from ether to photographic film.

It is by far the most valuable crop in the United States, and one of the four most valuable in the entire world. Some 67 million acres are grown in the United States each year, and the economies of entire states in the Middle West revolve around these waving green fields covering their plains. The average American annually consumes forty-five pounds of corn products directly and several hundred pounds of it indirectly through meat.

Other than an extreme energy shortage, it is hard to imagine any peacetime crisis which could bring the United States to her knees more quickly than a massive failure in the corn crop. But while other major nations have suffered agricultural calamities in recent years, few Americans can imagine it happening in their own country.

Yet in 1970 and 1971, the United States came very close to having exactly that situation.

The Famine That Knocked in 1970

In 1970, a disease commonly known as Southern corn leaf blight, which had long been a minor problem in a few isolated areas of the country, suddenly broke loose and invaded the entire corn-growing United States. In 1969, a new mutant of the disease had developed in the South, and in 1970, encouraged by a drought in most of the Corn Belt, it swept up through the United States and Canada. Eighty-five per cent of the corn grown in the United States were of hybrid varieties susceptible to the blight.

Crops were damaged, and the nation's corn output reduced. The corn market in the fall of 1970 fell into utter chaos, with prices skyrocketing from $1.00 to $1.60 a bushel. Beef prices were subsequently to begin an upward spiral which, aided by other factors, would continue for years.

A billion dollars in corn yields was lost, yet by November 1970, when Department of Agriculture officials and others looked back on the crisis, they were able to breathe a sigh of relief—it could easily have been worse. The corn blight, with its tremendous disruption of the agricultural economy, had effected a yield loss of only 15 per cent. The managers of the nation's bread basket had just experienced something closely akin to the tap on the shoulder a man experiences with a minor heart seizure. While 15 per cent is a significant figure to any farmer, it seemed small compared to what some agronomists feared for the following year; given the right (or, better said, wrong) weather conditions, there was every chance that a major portion of the U.S. corn crop could be destroyed by the leaf blight in 1971.

It was with an uneasy feeling that farmers began their planting in 1971. Most of them had been unable to obtain disease-resistant varieties of seed, and the fear of wholesale blight was running high. Anticipating lower yields, however, they planted more acres to corn, rather than less.

But several months before that planting season, an interesting alliance of scientific talent had begun to coalesce. As early as August 1970, NASA, Purdue University, and the University of Michigan had formed an ad hoc committee and begun flying aircraft over Indiana with various remote sensors, operating in a dozen wave lengths in visual and both near- and far-IR. They had then found that they received distinctive responses from blighted areas, in a certain narrow portion of the near-IR band; a lack of chlorophyll in diseased plants sharply reduced the energy radiated in that particular band. *The sick plants were withholding radiation, and the sensors were recording it.*

Combining their sensing with ground observations, the committee was able to classify signatures for various degrees of blight severity. Five classifications of corn condition were established, ranging from "1" (healthy) to "5" (severe blight). The tapes from the scanners had been fed into computers, which had printed out a map covered with numerals 1 to 5—indicating the degree of blight severity in each of the hundreds of plots under observation.

The results were intriguing. They showed that the blight (at least during its latter stages) could be detected from the air, much as it could be from the ground, and it could also be electronically mapped.

1971 Corn Blight Sensing

Botanists at the USDA were impressed, enough so that prior to the 1971 season they became part of the Corn Blight Watch committee,* formed to remotely sense the Corn Belt of the United States from high-altitude aircraft. By so doing, they thought perhaps they could map the fungus spread early enough so that selective spraying in infected areas could stop the blight in its tracks. Or, if that were not possible, perhaps farmers could at least be advised that their areas were on the way to being hopelessly invaded—allowing them to cut down their blighted crops, use the stalks as silage for cattle, and prevent the blight from moving on to other areas. If, early in the season, the blight was recognized as coming on with hopeless force, areas with planting seasons still to come could be advised to plant soybeans instead of corn; soybeans could fill some of the needs normally met by corn (such as feeding cattle) and reduce the potential calamity.

For observation, the committee selected no less than 210 sites, each one-by-eight miles, in Ohio, Illinois, Indiana, Missouri, Iowa, Minnesota, and Nebraska. Previous crop history of each of the plots, which covered 45,000 square miles in all, was compiled and put in computer banks.

Flights over the immense area were begun in April, and from June to October operated on a biweekly basis. The principal aircraft was NASA's RB-57, flying at 60,000 feet with color-IR film, sensing in the wave band found to be critical in the 1970 experiment. Three lower-altitude aircraft included the University of Michigan's C-47, equipped with a Multi-Spectral Scanner.

* With NASA, Purdue, and the University of Michigan. Numerous other agencies, federal, state, and local, ultimately were involved.

Ground observations were made to maintain a check on the performance of the sensors.

By June, the blight had begun to appear on the sensing data, which was being processed in Indiana with representatives from all seven corn states participating. Data was analyzed at Purdue and Michigan and forwarded to Washington.

Almost from the beginning, the preseason fears about the corn crop were given cause for relief. A warm spring allowed early planting, enabling the plants to have a healthy early growth prior to the blight season. Continued warmth in June favored more corn growth and gave the crop strength to ward off the first invasions of the blight. Then in July and August, when the usual hot, dry weather would have allowed blight to flourish, the weather trend reversed itself and cool, moist summer came on—the blight had been defeated at every turn by Mother Nature, and a national crisis never materialized. Instead, a bumper crop of corn was harvested.

A few members of the Corn Blight Watch were probably secretly disappointed that a major blight, plunging remote sensing into national prominence, had not occurred. But in spite of the low incidence of the blight, a tremendous experiment had been conducted. The data collected convinced a large segment of USDA that remote sensing could perform a vital service in monitoring potential national disasters. Techniques had been developed which could be quickly adapted to future crisis in corn or other crops, and the day of a satellite, as well as an airplane, being able to monitor a crop disease was now considerably closer.

In January 1972, *Successful Farming* magazine stated, "It is a scary situation when so much of our agriculture and food supply hinges on this one crop, corn. . . . With so much riding on the corn crop we need to set up a better monitoring system for detecting the threat of a disastrous outbreak, and the outbreak itself."

The monitoring system for large-scale control would seem unquestionably to be a satellite sensor.

Satellites to Give Food Planners More Options

If we look past our own borders, we can see even more emphatically that there is a lot "riding" on corn, and on other grain, and on the satellite which can watch the status of all these crops. The international aspects of the commodity market were underscored in 1972 when the United States made its first controversial grain exports to the Communist bloc. The world is full of countries growing rice, wheat, and other grains, all of which sometimes fail to produce, thereby creating potential customers for the United States.

There may well be occasions when it will be mutually advantageous for two countries both to learn that one of them has a crop failure ahead. A satellite may tell us in early spring that Mexico is doomed to a massive failure in her production of hybrid corn. Since the Corn Belt in the United States plants on a schedule several weeks later than Mexico, there would be time for midwestern farmers to switch some of their idle ground from soybean fields (a low-income crop) to corn, avoiding massive hunger in corn-consuming Mexico and increasing income for U.S. farmers.

As time goes on, the world or at least individual countries will be reaching decisions on how land and crops themselves are to be used: Will land be used to produce an export crop? Will it be used to supply food concentrates to the cattle industry? Or will it be used to produce foods for direct consumption by humans? Man being able to gain considerably more protein from the earth if he consumes vegetation directly, he might elect to eat sixteen pounds of corn instead of one pound of beef, a policy which might not excite the affluent but would perhaps appeal mightily to the country concerned about feeding an exploded population.

Our interest in such decisions is that more of these options can be exercised if careful satellite crop monitoring is conducted. We have already noted how planners can control a crop to meet as many needs as possible, as in the raisin case. Likewise, by monitoring the health of their fields, they can avoid famines, or at the very least foresee them, and make emergency plans—such as slaughtering calves early to retain corn for human consumption.

Ideally, however, satellite information will be used not only to forecast shortages but also to avoid them. Just as the raisin supply was controlled from the air, so are satellites gradually being used to better utilize rangeland or cropland of any type. In California, some agriculturists foresee a day when satellite information will be sophisticated enough so that growers of a given crop can plan in advance for activities of both their own operations and related industry. They anticipate satellite information for example, which would enable tomato growers in various valleys to coordinate in advance the flow of laborers, cartons, trucks, and railroad cars for their respective harvests. The proportion of tomatoes which normally perish because of lack of planning information could be reduced. Packing houses, carriers, and marketing organizations could prepare their labor and distribution systems in advance. The life and plans of migrant farm workers could be more efficiently and comfortably arranged. And advance knowledge of blights and insect infestations (determined by a sophisticated sort of corn blight watch) could advise farmers when chemical usage would be necessary. All of this depends on continued refinement of computer interpretation. And many agriculturists specializing in sensing are now optimistic that data interpretation will develop farm planning spectacularly.

Satellite War Against Pests

In the case of insects, a rather impressive example of advance planning has already occurred. In the fall of 1972, University of California research scientist Virginia Coleman stood in an Imperial Valley cotton field and felt one of the fluffy white bolls on the bare stalks. The cotton stalks already had been stripped of their leaves by chemical defoliant. Now, at harvest time, huge whirring cotton pickers were noisily wheeling down the rows, plucking off the balls of fluff and throwing them into the screened hoppers and machines carried on their backs.

The balls of cotton piling up on the hoppers looked like cumulus clouds building up into a gigantic thunderhead, but the pileup of the fluff, however dramatic, was slower than in years past. Several years ago, the field where Virginia now stood had produced almost two bales per acre. This year it was down to one bale. The difference lay in the activities of the boll worm, a destructive caterpillar with a habit of boring into a cotton boll and opening a tunnel, through which an army of other insects could then pass. Each boll so selected by the caterpillar was destined to be destroyed weeks before harvest could take place.

By January 1973, the whirring mechanical pickers had plucked off what bolls there were in the Imperial Valley, and the cotton stalks then stood completely naked in the winter sun. Now that the boll heyday was over, the worms had nowhere to go but to burrow into the stalks, making off-season homes, and spinning cocoons in which to wait out the winter.

However, the California Department of Agriculture had other ideas for the boll worms. All farmers were required to plow their cotton stalks under ground no later than January 15, in an effort to destroy the "host" environment of the worms. Once a stalk was plowed deep into the ground, the worm's chances of maturing into a moth and mother were markedly reduced. But it was important that all the farmers comply; otherwise one bad field, like one bad apple, could in time "spoil the barrel."

On January 17, ERTS passed over the Imperial Valley, as did a NASA U-2 aircraft, and Virginia Coleman awaited the imagery that would be sent to her. Throughout the autumn she had studied the ERTS pictures, identifying all of the dozens of fruit and vegetable crops of the valley, and particularly the cotton fields. Then after the bushy green plants had been defoliated into barren stalks, the near-IR response had dropped sharply in successive imagery; ERTS was giving her a marked distinction between the earlier bushy fields and the later stalks.

The purpose in January was to go one step further and try to differentiate between fields with stalks and those that had been plowed under as prescribed by law. The agricultural authorities each year had attempted to hunt down violators of the plow-down regulation, but their methods had proved time-consuming and ineffective. If the satellite, or at least the U-2, could provide a map of violations, real progress might be made. Virginia Coleman was anxious to see the imagery.

When the pictures arrived, she was not disappointed. She found that ERTS, as well as U-2 imagery, was satisfactory. "There were decided differences between the January ERTS pictures of plowed fields and the earlier images of fields with the bare stalks standing," she said. The pictures were made available as evidence for agriculture officials to use in court.

More important, "once the 'bad apple' farmers see there are some teeth in the plow-down regulations, everyone should begin to comply," Miss Coleman predicted.

The chances of picking two bales of cotton per acre had suddenly become a realistic hope again. Not only could the food (or, in this case, fiber) output of the land double, but the chances of preventing the boll worm from spreading to uninfected areas looked brighter. A satellite 570 miles away had proved it could be used to control a war against an insect the size of Virginia Coleman's little finger.

In the case of the Imperial Valley, with only 80,000 acres of cotton, the U-2 was a practical sensing machine, and ERTS, frankly, was not needed. However, if the study were expanded to cover just the state of California, with ten times the Imperial Valley's cotton acreage, the cost of the U-2 would be prohibitive.

Or would it? For what volume of sensing would aircraft data be useful, and how much acreage would it take before satellite expense would be justified? Was it time for a cost study?

Even before the ERTS-1 reports were all collected, Miss Coleman was proposing an advanced boll-worm study for ERTS-2. She intended to find ways of using satellite data effectively and combining it with aircraft and ground information to gain the best features of all three. Such a project would have to be cost-analyzed, so she would have to find a financial person to join her study team. It would soon be routine, she thought, for agencies to look at satellite sensing with an accountant's critical eye. Cost-benefit studies should rapidly become a part of remote sensing, once the first bloom of earth observations uniqueness was past.

"We have to find where satellites are practical and where they aren't," Virginia Coleman said. "There's no sense in doing anything with a satellite permanently unless it does a unique job, or unless it costs less than other research—one or the other. And preferably both."

Save satellites for the tasks where they are needed—a logical rule, and one which hopefully will be followed. Satellites will obviously prove to be the means for broad-area tasks, with airplanes used for localized work; and studies such as Miss Coleman's ERTS-2 project will eventually determine at what point one is more economical than the other. The satellite and the aircraft should be complementary, rather than competitive, and there is considerable need for both of them in this world where no one can accurately estimate the food production, where valuable cropland lies unrecognized, and where there are such variable factors as rainfall, blight, and insects.

9. Water the Vital Ingredient

Historically, man has reacted to the great volume of water in this world of ours, the solar system's remarkable "wet planet," as more of an obstacle than a resource.

Small wonder. Some 97.3 per cent of the world's water is contained in its immense oceans and is generously seasoned with salt. The principal role of this major portion of our water, as any history book will show, has been to keep us from moving freely over the earth. Nevertheless, we have managed to navigate across it (at speeds we would consider ridiculous ashore) and, more recently, to fly over it, but all of this in order to reach the more precious commodity, land.

We've fished in it, of course, and a few maritime nations have substantially stocked their larders with its morsels. But on a worldwide basis, we have used it to provide less than 5 per cent of our protein. We have seldom drunk it or lived in it, or used its vast power to provide energy. Actually, our principal use of the world's sea water has been to dilute the wastes we have poured off the edges of our continents.

Of the 2.7 per cent of the world's water that is salt-free, most of that (2.2 per cent[1]) is at the poles frozen into great chunks of ice. These have served mainly as burdensome fences, blocking us from the otherwise convenient travel lanes across the Arctic. Occasionally they have even floated away from the poles and collided with our ships.

Thus we account for 99.5 per cent of the earth's moisture as being either sea water or ice, left to us by Nature as an option to venture into or ignore as we choose.

That leaves just half of 1 per cent, which has been managed for us and delivered to us with reasonable regularity by Nature. *She* has been the one to find use for the oceans, continuously heating the surface of the water, distilling it, and then storing the desalinized moisture in her skies, eventually to carry it to us as rainfall. (In addition, she long ago set aside a sizable portion of fresh water in a trust account for us in lakes, glaciers, and underground aquifers.)

Now, half a per cent is a modest enough share of anything, but when we are talking about consuming only that fraction of a renewable resource, it is minuscule indeed. And yet, while the dehydrated folk of the Mojave Desert would like our rainfall to be greater, the residents of the Amazon Basin undoubtedly think it more than ample. The truth is that half a per cent could provide a comfortable global water supply, but man has done a poor job of utilizing and distributing it.

Man has divided that half per cent of the world's water—our rainfall—into two portions, using one and discarding the other. Referring first to the portion we use of our U.S. rainfall, we consume only 7 per cent in our municipalities, our industry, and our irrigated crops. Another 23 per cent of the rain is put to good use by Nature, through no effort of man, by being absorbed on nonirrigated farms and rangelands on which it falls. Approximately 16 per cent provides similar moisture for forests and wilderness areas.

On the other hand, that portion of our rainfall which we do *not* use includes some 22 per cent which is returned downstream to the ocean unconsumed, albeit used to dilute our sewage and run a few generators along the way. The largest portion of all, however, a startling 32 per cent, simply evaporates, or evapotranspirates into the air through plants.[2]

By thus letting about half of the U.S. rainfall slip through our fingers, we are to a large degree to blame for having the use of only a fourth of a per cent of Nature's water. This reduced percentage seems to reflect an absurd waste, when we realize that we can expect a large section of the United States to have drought conditions every year—or when we note that we live in a world largely covered with water while its people often do without food. Water and soil simply haven't been brought together.

We have observed in previous chapters how satellites will help us find land and determine its potential use; through this process, we will almost certainly find billions of acres of idle land that could, with the addition of water, become cultivable. The same satellite that so masterfully finds fertile but parched land for us could deliver a *pièce de résistance* by telling us how to obtain water, the missing vital ingredient. Or, better yet, it could help us evaluate not one but several water sources. In this chapter, we will examine half a dozen such options.

Diverting the Water to Where the People Are

In the United States west of the Mississippi, water is in short supply more places than not, although there are some striking exceptions. The dry, windy Dust Bowl of today covers even a greater area than it did in Steinbeck's *Grapes of Wrath;* it extends from Texas to the southern tip of South Dakota.

But not so in the Pacific Northwest. Dust is, in fact, hard to find there, in an environment where Nature is disposed to rinse things off every few hours, creating a green, if somewhat overcast, atrium between the Pacific Ocean and the Cascade Mountains (1973 was a remarkable exception). It is not surprising that senators and congressmen from Washington and Oregon quickly banded together in 1968 when there was talk in Congress of redistributing western water from the haves to the have-nots. The concept of redistributing water, however, is one of the options at which man is taking a long look, even though the Northwest lawmakers would prefer that the idea be permanently forgotten.

The wariness among Oregonians and Washingtonians is based in part on the goings on they have observed in the unpredictable state to their south. California, for three quarters of a century, has been constructing a gigantic statewide network of dams and canals, redistributing water from the wet lesser-populated areas in the Sierras and the Sacramento Valley to the dry coastal areas. Aptly labeled the "Aqueduct Empire" in a book by Erwin Cooper,[3] the waterways have been built, with Roman-like intensity, by government sponsors ranging from local to federal.

The now fabulously productive Imperial Valley was built with water diverted from the Colorado River in 1903. Shortly thereafter, San Francisco and the dry, but already populous, Los Angeles quickly caught on to the Roman technique, each of these cities electing to pipe in mountain water from hundreds of miles away. In the 1930s the U. S. Bureau of Reclamation's Central Valley Project, a vast network of dams and canals, began diverting water from the frequently flooded Sacramento Valley to the always thirsty San Joaquin Valley.

The floodgates had barely been opened on the federal CVP when the state of California, not to be outdone by the feds, began building its own immense system of waterways in the 1960s. It included the highest dam in the United States (the Oroville), the widest dam in the world (the San Luis), and an aqueduct which orbiting Apollo astronauts would be able to see with their naked eyes.* Some four million acre-feet of water is annually transported for distances as great as seven hundred miles. About a million lush acres in the San Joaquin Valley is irrigated as a result. The amount of water to be diverted is to be gradually increased, according to need, until 1990, with subsequent canals to be built in the meantime.

However, all of this diversion of water from northern California rivers has drastically reduced the flow from the rivers into San Francisco Bay, causing argument that the bay might become a stagnant, salty cesspool. The critics also maintain that the decrease in fresh water is going to allow more salt water to flow upstream into the California delta than can be effectively pumped out, thus destroying what is probably the most fertile farming area in the state. Furthermore, they argue, the salt water will flow upstream several miles *under* the fresh

* The only other man-made feature they could see was the China Wall.

water. The fish and other creatures that dwell in the river will then have only a thin surface layer of fresh water, a nutrient trap in which oxygen will be depleted and the rivers periodically turned into a mass of decaying algae. Conservationists already complain that the changing bay is becoming devoid of fish and birdlife.

Bay area residents have pondered these possible dilemmas with increasing annoyance. The pipes of the California Water Plan, which in the dry San Joaquin areas were seen as a rancher's dream, had in San Francisco taken on the image of a plumber's nightmare.

The concern of bay area residents lay not only in their present troubled waters, but also in the fact that the Oroville dam was scheduled to divert more and more water, gradually to accentuate the bay condition as time ticked on toward 1990. And the same worried people have become even more apprehensive of the Omnibus Water Bill which U. S. Congressman Bernard Sisk of Fresno introduced in 1973, and plans to reintroduce in 1975. Sisk's bill calls for yet another massive federal project, annually diverting 1.5 million acre-feet of additional water to the San Joaquin Valley from three northern tributary rivers whose water presently winds up in the bay.

How would Sisk's bill affect the bay? An understanding of the bay's present condition is the first step in answering that question. Such an understanding can be partly gained through ERTS and aircraft sensors being utilized by the marine geologists of the U. S. Geological Survey.

Beginning in 1969, a USGS study team began observing and photographing the sediment pattern from low-level airplane flights. The Sacramento River drains into the northern end of the bay. During the dry summer, the river water meekly takes the most direct path possible through the northern bay and rolls gently out to sea under the Golden Gate. During the winter and spring, however, the Sacramento water surges southward throughout much of the bay, churning around the shoreline before finally flowing seaward. In the process, it flushes out the bay and many of the pollutants in it.

This phenomenon was noticeable in the USGS aerial pictures, the muddier fresh water being distinguishable from the regular bay water because of its brownish turbidity. (Ground truth, or, in this case, *water* truth, was gathered by boats and subsurface devices, confirming the fact that the aerial pictures indeed were accurately distinguishing between fresh and salt waters.) The pictures offered good basic information, and they were studied by people on both sides of the water controversy. They also offered a data base for USGS marine geologists, who prepared to expand their study with ERTS pictures of the bay waters.

The first several months of ERTS sensing, unfortunately, were blocked by cloudy bay area skies, but by early 1973 ERTS was able to provide clear pictures of the water. The composite ERTS pictures, including all four bands, showed little if any sign of sediment. However, when marine geologist Dr. Paul Carlson viewed transparencies from each of the four bands separately, he was able to see the sediment clearly in the green band. The same sediment pattern was seen on one occasion in U-2 pictures of the bay (taken almost simultaneously with ERTS), and was further substantiated by ground-truth samples secured by boats. ERTS's credentials on the job were now established.

Subsequent ERTS images were then used with confidence throughout 1973, mapping the sediment pattern of the bay at different seasonal river levels. A seasonal pattern for the action of the bay's fresh and salt water was thus established. Hydrologists should now be able to project what changes would occur with the even lower water levels anticipated leading up to and after 1990. The same kind of projection could be made for the delta upstream, forecasting what situation would be created there with increased salt-water intrusion.

Each time the California Water Resources Control Board meets, it should now have a portfolio of ERTS and related documentation to review, along with expected arguments from San Joaquin farmers and ranchers on the one hand, bay area residents and Sierra Club members on the other. Should the gigantic water project be expanded to irrigate more farms? Or would the bay suffer more than the farmers gained? Thanks to ERTS, the board will be able to base its decision, at least to some degree, on hydrological projection, rather than entirely on politically flavored speculation.

If the ERTS and other data should lead the board to a negative decision, what then? How will southern California overcome its problems of drought and expand its croplands into the deserts? If ERTS shall have scuttled one option, it seems only fitting and fair that we look to it for an alternate plan. Appropriately, several investigators from California and other western states are dedicated to this purpose, intensely combing ERTS data for sources of water.

Saving the Lakes That Evaporate Away

Three years before ERTS, however, a remarkable photo of the Texas–New Mexico border was brought back from the productive Apollo 9 flight. For anyone who has been assured since early childhood that the artificial boundaries drawn on maps do not exist in real life, the Apollo photo is a revelation. The Texas–New Mexico border, in this unusual picture, has just as distinct a line as it would on a map in an atlas. The marked contrast is a result of Apollo's near-IR bands, showing dense red vegetation on the Texas side of the border, juxtaposed against more subdued color across the state line.

The reason for the difference in vegetation, according to Dr. C. C. Reeves, hydrologist at Texas Tech University, may be found in the water regulations of the two states. The Texas laws for drilling wells are considerably more liberal, resulting in a flourishing growth of cotton and milo maize that continues right up to the border and stops.

In supporting the necessary irrigation for this growth, Texans had lavishly drilled more than 66,000 wells in the southwestern High Plains area by the end of 1972. The consensus was that they tapped the water table to a maximum. In some parts of the district, the level of the water table had dropped markedly.

"Things are catching up with us," Reeves said. "We're going to have to make a choice between more water and fewer people."

"Fewer people" being a difficult policy to invoke anywhere, the Texans have studied two possible plans for increasing the water supply. The most grandiose was a system to transport water uphill from the Mississippi River, some seven hundred miles to the east. This California-style plan would have been developed along characteristic Texas proportions, requiring a gargantuan pipeline and pumping operation. It was defeated at the polls in 1968, probably because of its extravagant potential cost, leaving West Texans to search for another source.

They found one. Dry as West Texas is most of the year, it is subject to cloudbursts and heavy rains each spring and fall. Following these deluges, thousands of lakes are formed, most of them destined to evaporate quickly in the following few months.†

This water wasted by evaporation is considerably greater than the total water consumed in the area, and it creates an obvious challenge. How could this lake water somehow be captured before it would drift off into the sky twice a year?

The solution Texans are proposing in answer to this question is an old but hydrologically sound one. They would pump the water into the ground, sealing off the opportunity for evaporation. With this system they could recharge the water table as much as they wanted, or at least as much as they could afford to pump.

This kind of project would be costly, and its over-all feasibility hinges on determining the distribution of the water in the thousands of lakes. Just where would pumps prove valuable, and where would they turn out to be white elephants, with relatively infrequent water to pump?

Up until now, no one has developed a technique to estimate the volume of lake water held in the various basins of the Plains area. The best estimates prior to ERTS had "narrowed it down" to a range of from 1.8 to 5.7 million acre-feet, hardly a

† Annual rainfall of 19 inches compares with an annual evaporation rate of 60 inches, an overdrying of three to one.

pinpoint appraisal.[4] In 1972, Reeves became chief investigator on an ERTS study designed to map the standing water in the area. The ERTS data would provide a regular (eighteen-day) picture of how much land was covered by water; along with this, ground surveys in selected basins were conducted to convert the two-dimensional satellite pictures into cubic water measurements.

By the fall of 1973, after a successful year with ERTS, Reeves commented that he felt it had been "conclusively proved that water census by satellite could be effective." Meanwhile, the Army Corps of Engineers had made considerable progress in developing an accurate means of converting the data to cubic measurements. The Texas Water Development Board was so highly impressed with the results that a permanent program was established to expand Reeves's work on a statewide basis. Wherever the evaporation of lake water coincides with a water-short region in Texas, the practice of recharging is being evaluated. When the survey is complete, the state will have a cost-benefit ratio for recharging systems; as Reeves puts it, "ERTS is providing the first step in what might truly become a perpetual water supply."

Search for Water under the Desert

Recharging the ground may work for West Texas, but if we go westward across the southwestern desert, we find less and less rain to be used or stored in any form, underground or otherwise. The irrigated farms which decreased sharply at the Texas border we find replaced by arid rangeland in New Mexico, and finally, in the drier reaches of Arizona and southern California, by sheer desert. Underneath the dry sand there are layers of water, but man has found by experience that wells in such areas last for a few years and then give out. Farmers drill their pipes several feet deeper into the sand each year until, as with a youngster sucking the last of his soda from a straw, the pipes gasp and the water sinks out of reach.

The large agricultural farm regions, such as Phoenix and the Imperial Valley, exist from river water, but in between these lush areas there are only scattered small oases, where wells are flourishing, presumably none of them for very long. Tens of thousands of wells have been drilled in the Southwest, but with no truly long-term water supply having ever been tapped.

Perhaps ERTS can help discover one.

There are now numerous people (not all of them formally working with NASA or on ERTS) looking at MSS imagery to search for different sources of desert moisture. In the University of California at Riverside, ERTS data is being studied by graduate students intending to map those areas where yucca is growing in the desert. Yucca being a shallow-rooted plant, the purpose of the study is not to locate potential well sites, but to use yucca stands as a means of determining areas where *other* low-water crops might flourish, without irrigation and for a prolonged period.

A more dramatic program is studying geological structures in the southwestern states. The project includes study of data from Skylab as well as from ERTS. Its primary purpose is earthquake-fault study, but investigators had the foresight to list water-prospecting as a secondary mission; there is no reason a geologist cannot look at fault structures for earthquake research and simultaneously keep alert for areas holding high probability of minerals, oil, or water deposits.

Geologist Dr. Monem Abdel-Gawad has reason to believe that ancient Arizona had a system of rivers, flowing southward, bearing absolutely no relationship to the northbound river pattern that exists there today. ERTS is providing the first regional geological view of the Southwest and will hopefully reveal traces of the ancient streams. Abdel-Gawad believes that those rivers, like the desert streams of today, had underground springs proceeding directly underneath the beds of the surface streams. Eons ago, when the faulting and tilting known to have occurred in the Southwest took place, drainage would have been abruptly reversed, with immense pools of subsurface water trapped as a result, Abdel-Gawad believes.

This kind of pool would probably be capable of yielding water that would burst out of the ground at several times the speed of well water pumped from the surrounding water table. The validity of the theory seems plausible when we look at a huge encasement of water discovered, hundreds of feet down, in structures near the Imperial Valley. Water there is capable of streaming to the surface, at speeds that approach the speed of sound, under a pressure of 2,000 pounds per square inch. (Alas, these Imperial Valley cisterns yield not clear cool water, but hot salt water. This salt water comes as no great surprise to geologists familiar with the history of the area; in fact, they find it an encouraging sign that underground rivers of the same general period may indeed be similarly entombed, offering untapped fresh-water reservoirs.)

Water Parasites

Enough said about the capricious underground. While it's true that there is often available water hidden beneath our very feet, there are also millions of acre-feet of moisture openly going to waste every year in the surface riverbeds of our western states.

On the Gila River in Arizona, the U. S. Geological Survey has spent several years experimenting with salt cedar plants which grow along the river. Salt cedars are plants with a voracious thirst; they consume immense quantities of water through their roots, and evapotranspirate it into the atmosphere almost as fast as an incinerator turns fuel into stack gas. Experimenters have eliminated salt cedar along selected stretches of a test area on banks of the Gila and substituted grass for it. By studying the river's flow in both the salt-cedar and the grass areas, the experimenters have been able to determine how much water is consumed by the river plants. Their conclusion is that the salt cedar in the test area is consuming no less than *two acre-feet of water per year.*

Since the mid-1960s, the USGS has monitored the study area with aerial photography, developing signatures for the salt cedars. Using these signatures, they have since utilized ERTS data over most of Arizona and located salt-cedar populations along thousands of miles of streams. The amount of water they consume is variable.

"To determine how 'thirsty' the plants are in any particular area," says Richard C. Culler, Department of the Interior's Water Resources Division, "we use a formula based on temperature, climate factors, and plant density." ERTS is thus enabling the study group not only to identify salt-cedar areas but also to determine the amount of water being siphoned out of a given streambed. "Any southwestern area needing water could thus use ERTS pictures to determine how much water could be made available by eliminating the plants."

The formula reveals that salt cedars in various situations use water at rates *ranging from one to nine acre-feet per year.* Calculating the salt-cedar population in the Southwest, one Interior Department spokesman conservatively estimates that 16 million acre-feet could be saved annually—three times as much as Arizona now uses to irrigate! The water saved would have a dollar value well in excess of a billion dollars a year. Salt cedars, in fact, have little if any economic value, whereas grass does. In the case of the Gila test area, the grass planted was used to graze cattle belonging to the Apache Indians.

We should interject that salt cedar is not without its staunch supporters. The plants do offer a tall shelter for wildlife, providing the sort of atmosphere which outdoorsmen like to see maintained along meandering rivers. This kind of aesthetic interest versus such a bread-and-butter motivation as water availability creates controversies to be resolved, not in a satellite system, but in the time-honored halls of debate. What ERTS does is offer man the total scope of the problem to examine in making his decision. Without satellites, the decision makers would not be aware of the immense volume of water being evapotranspirated.

Sweetening the Sea Water

The water lost through salt cedar is not the only highly visible potential water source. If any single concept has been continuously viewed as the panacea for water problems everywhere, it has been desalinization. Its fulfullment has been the charge laid upon national and state task forces in the United States, and was to have been the principal thrust of the International Hydrological Decade, now winding down into termination in 1975. The search for "sweet water" may predate history, and there is evidence of men looking for it as early as 3500 B.C.

Desalinization is a daily fact of life, as any sailor knows, and it is accomplished ashore as well as at sea. When Fidel Castro shut off the U.S. water in Guantánamo Bay in 1964, the Navy was there, waiting, with a distillation plant. This should not have surprised Fidel; several small communities in the world had existed on distilled water for prolonged periods, albeit at great expense.

And therein lies the problem. We could talk about dozens of desalinization plans which bloomed in an inventor's eye but withered at the comptroller's desk. The cost of any proved desalination method still runs at several hundred dollars an acre-foot, many times the cost of even California's relatively expensive ($50 to $60 per acre-foot) water.

Still, technology has a way of shattering ancient problems overnight, and this one should ultimately be no exception. The Atomic Energy Commission is attracted to the concept of nuclear systems producing desalinized water and electrical energy simultaneously, and such plants may begin dotting our shores in the next few years.

At that point, earth satellites will have a significant job to do. Almost all the attempts at saline conversion have involved considerable amount of heating of water. Distillation, copying nature's own desalinization system, is the most frequently mentioned concept, and it requires heating the sea water and then cooling the salt-free steam into fresh water. If such a process became popular worldwide, the amount of thermal change in the lakes, and rivers, and conceivably even the oceans of the earth could be great. But how great? Some scientists say we would eventually have all the waterways of the world (including the oceans) thermally polluted, with such ultimate disaster as melting of icecaps and other nightmares. Other equally competent scientists emphatically disagree. It would obviously be an advantage to *know* the extent and effect of thermal pollution in any system ahead of time; national and ultimately global decisions would likely have to be made regarding regulation and control.

Well in advance of desalinization plants, weather satellites (see Chapter 11) will hopefully have become engaged in the monitoring of water surface temperatures around our present nuclear power plants. Models determining the degree and distance of thermal pollution will presumably have been developed from the temperature findings of the satellites. With such models, scientists can begin to calculate how the temperature changes will affect marine life, and icecaps, and all of Nature. Then man can make his decisions. Without such models, we will be operating on a trial-and-error basis.

Addressing himself to the matter of potential worldwide thermal pollution, ecologist Barry Commoner has said, "We are conducting a huge experiment on ourselves."[5] Satellites offer a chance to bring some light on the subject while the experiment is still small, rather than waiting until the world has become one vast global laboratory.

The International Water Supply in Antarctica

In looking at the world from space, it becomes difficult to alter the forces of Nature concerning one nation without also affecting the inhabitants of another. The problem of thermal pollution of oceans, if in fact it *is* a great problem, is not the only global experience in which the nations of the world are participating. The whole subject of water availability is one which exists worldwide (and to a considerably more acute degree in those parts of the world where persons exist on a few quarts a day, as compared to a U.S. per capita consumption of hundreds of gallons daily).

If we were to leave the United States and meander southward in our own hemisphere, we would alternately pass areas of rain forest and drought—first in Mexico, then Central America, and then on both sides of the Andes. Colombia has recorded 28 feet of rainfall in a single year, and yet, on nearly the same parallel, in northeastern Brazil, the dry years of the *seccas* are a sad fact of life.

In northern Chile, the warm, fertile fields are barren for lack of rain, with the northern Chilean being much like a man dying of thirst with an ice tray in the next room. The "ice tray" is the continent of Antarctica, where 95 per cent of the world's ice exists and goes unused. To some scientists, Antarctica seems a logical source of water, not only for Chile but for the entire world. Each year, large pieces of ice break off the Antarctic continent and become icebergs. The bergs are long and thin, their proportions almost like ice shavings; their width is almost inevitably less than a mile, with their lengths ranging from one to a hundred miles. The idea of securing these bergs, towing them to arid parts of the world, and melting them, has seemed practical to a number of physicists and engineers for years.

The RAND Corporation, the nonprofit think tank in Santa Monica, California, would like to see the iceberg concept initiated in an experiment for Los Angeles. They visualize towing bergs from Antarctica at a slow, low-friction speed and docking them a few miles off the California coast. There the ice would be quarried into chunks, conveyed to a tower, and fed into underwater pipes to be carried ashore. Dr. John Hult of RAND estimates that the water could be delivered from Antarctica into the California water system at a cost of $25 per acre-foot, much less than present water costs there.

"A square-mile iceberg, six hundred feet deep, could probably supply Los Angeles with all its fresh-water needs for six months," Hult estimates. *Another potential which Hult visualizes is the ability of the ice to absorb thermal ocean pollution from power plants, nuclear or otherwise.*

The iceberg concept has been popular among scientists for a third of a century. Probably the major obstacle which has precluded action has been the formidable Antarctic itself. Man has found visiting Antarctica a slow, frigid, often perilous, and always expensive venture. For ships, the icebergs have proved a major navigational hazard; for aircraft, there has been the ice blindness, the eccentricity of compasses near the pole, and the dangers of operating from solid-ice runways. Hence man has not even been able to get to the first stage of surveying the iceberg resources available for transport. For example, the selection of proper icebergs for towing is one critical phase of the project; and most of the bergs sighted by Russian surface ships, in their necessarily cautious voyages, have been ruled out by physicists as too small to be towed efficiently. The size and shape of an iceberg are factors related to such economics as the speed at which the berg can be towed, the time it will take to melt, and the sea lanes which can be followed in towing it to its destination.

RAND's completed ERTS-1 project and proposed ERTS-2 project are now offering man a chance to begin locating the icebergs, determining their characteristics, and charting routes to reach them and tow them away. ERTS-1 provided imagery of the berg areas during the Antarctic summer of October 1972 to March 1973. Dr. Hult was then able to obtain detailed measurements of numerous suitable bergs in various areas, and was enthusiastic about the role of ERTS. "We're convinced now that satellites offer the only means of surveying the Antarctic," he said. With ERTS-2 Hult expects to accumulate a wealth of additional information, information critical in managing an iceberg expedition all the way from the designing of towing equipment through the actual conducting of a safe operation.

Barring political obstacles, the U.S. experimental ice-towing program would seem to have an excellent chance of being initiated in the late 1970s. When this does occur, the Antarctic will have been dramatically unlocked by a key turned from 570 miles overhead.

Flexibility Provided by ERTS

ERTS has provided the iceberg data in some of the same orbits in which she took pictures of the Arizona rivers and the San Francisco Bay water, allowing man simultaneously *to weigh options which might otherwise have come to his attention a hundred years apart.* Solutions which might seem wise, if discovered one at a time, may be recognized as shortsighted when compared side by side with other options. Or two options may be recognized as complementary. Combining a cold iceberg with a hot nuclear plant in the same coastal waters might neutralize objections to both concepts.

Whatever options are exercised, water is certain to flow into areas where it will be the missing ingredient for opening up croplands and filling other vital needs. The Brazilian family which exists on limited water, the children who die hungry in countries where fertile land stands parched and idle, and the American cities which may face food shortages all stand to benefit.

10. Ocean Harvest

Oceanology has just never gotten off the ground.

In the 1960s, much was said about undersea hotels, submersible seacraft, deep-water recreation, and the extraction of drugs from the ocean depths. A national news magazine indicated that submarines would soon scour the ocean bottoms in search of everything from sunken galleons to mineral outcrops. A hungry world could begin to feed itself from the seemingly inexhaustible food supply of the sea. (As a beginning, the United States might dike the Pacific coastal tidelands and harvest fish by the acre much as farmers do corn.)

Concepts mushroomed. We seemed to be embarking on a venture that deserved a new name, a new federal agency, and a grandiose new budget. The term "oceanography" was too narrow to encompass all of man's proposed forays into the sea, only some of which would be in the name of science; so a word was coined for the new technology, "oceanology." A new federal agency, patterned after NASA, was advocated in 1969, aimed at making the 1970s almost as dedicated to the oceans as the 1960s had been to space; $36 billion had been spent on the space decade and $20 billion was suggested as a ten-year budget for the "wet NASA." And on that proposal ended the decade.

Once again, however, as in space, the United States was to cut back, reassess its direction, and decide to spend less of its money on exploring the unknown. The "wet NASA" was created, true enough, in 1970; it became the National Oceanic and Atmospheric Agency (NOAA, pronounced "Noah" and appropriately responsible for matters of water and weather). The agency was not to be independent, however, as NASA was, but would be assigned to the Department of Commerce. It absorbed more than a dozen federal entities, most of them oriented to either weather or fishing. Its budget in 1973 was just under $400 million, a rate which, if continued, would mean a more modest $4 billion rather than the lavish $20 billion decade.

When the braking action by the federal government occurred, the movement in the private sector sputtered as well. Submersible seacraft went into mothballs, the architects for underwater buildings put away their drawings, the recreation industry never went beyond scuba diving, and sea mining remained experimental. There was no industrial giant standing in the wings, willing to fill the breach; no one ready to drop several billion dollars of bait into Davy Jones's locker, or willing to work twenty years to reel in the rewards.

Momentum was lost, but the challenges of the sea were still there. With the large submersibles out of the picture, more economical concepts were needed. On a less ostentatious scale, the research of the ocean would move ahead, but with priorities established. Although private enterprise (including Howard Hughes) and some foreign governments would secretly develop undersea mining, the big push at NOAA, and throughout the world, would surely be toward the search for ocean food.

Hunting and Farming at Sea

Much was being said about the concepts of "hunting" food in the sea versus "farming" it. Virtually all of today's seafood is still a result of men going down to the sea with nets and poles. We had, of course, long ago domesticated the animals and plants of the land, and oceanologists were discontent that we were still chasing fish around the way we had once hunted buffalo and plucked wild berries. We had farmed shellfish for several millennia, but were just getting around to that point with fin fish. If you define farming as the culturing of fish all the way from egg to harvest, the British of the 1960s were the very first to achieve it. The British achievement came on as a smashing success, although several problems prevented it from springing into large-scale practice. Japan, meanwhile, continued improving its culturing of oysters and other shellfish. And the Soviet Union made progress in fish culturing while continuing to be a worldwide hunter.

But whether hunting or farming, our goal everywhere was the same: more protein for a world that was finding it difficult to feed itself from the land. Japan was forced to obtain almost two thirds of its protein from the sea. Peru, which had emerged as a new maritime nation in the early 1960s with the highest catch of any nation, had significantly supplied low-cost fishmeal to the U.S. poultry and European cattle industries. The sea seemed to be burgeoning.

Food Chain Begins with Sunlight

The concept that the sea was a great cornucopia of world food was tempered, however, by studies by the U.S. government, by UNESCO, and by simple statements of fact by oceanographers. The oft-heralded "inexhaustible" food supply began to look like a myth when facts were examined; in spite of the great depths of the oceans, we were reminded that all life in the sea, as on land, stems from the sun's rays. Photosynthesis occurs only in the shallow photic layer of the ocean, that top hundred feet or so where sunlight streams in and is captured by the cells of phytoplankton and seaweed. This plant life is the first link in the ocean's food chain. And not unlike land dwellers, the sea creatures tend to congregate near the kitchen. The zooplankton and sardines stay near the plant life they consume, the herring near the sardine it eats, and for similar reasons the mackerel near the herring, the tuna near the mackerel, the cod near the tuna, and the shark near them all. There are relatively few true denizens of the deep—lantern fish and the like living continuously in the dark, lower levels of the sea—and even they owe their existence to the photic layer, eating manna raining down on them (in the form of table scraps from the chain of life above). All life in the sea is thus a product of the surface waters and a very finite thing.*

If we were to use algae as a food for ourselves, the protein potential of the ocean would indeed be bountiful. (Pulverizing algae could conceivably be the trick that would solve the global food problem, but up till now algae has proved to be as indigestible as it is distasteful. Adding insult to injury, dried algae has cost twenty times as much as our more expensive fish.) For the foreseeable future, at least, we will be doing our eating a little further down on the food chain, where, unfortunately, the volume of food available to us is sharply reduced. The first stage in the food chain effects only about 80 per cent reduction, indicating that a ton of algae would yield a full 400 pounds of sardines—rather efficient, actually. But, as we move further along the chain, a larger percentage of food is lost.[1] Thus the 400 pounds of sardines become 40 pounds of herring, then 4 pounds of mackerel, 6 ounces of tuna, and less than an ounce of cod.

By contrast, in the shorter food chain on land, a ton of corn produces 250 pounds of beef.

After allowing for the inedible portions of both the beef calf and the cod, we calculate the following: A person gorging himself on a 16-ounce chunk of sirloin is, indirectly, consuming 14 pounds of corn; whereas a more restrained eater, confining himself to a modest 6-ounce cod steak, is taking on no less than 40 *tons* of seaweed.

The food chains of the sea being as uneconomical as they are, the pessimistic findings by the President's Scientific Advisory Committee (1967) are not surprising. Scientists on the committee had varying ideas, but all fell considerably short of projecting an "inexhaustible sea." They variously estimated the potential harvest of the ocean at totals ranging from only double to eight times its present yield.

* The "red tide" experienced by New Englanders in September 1972 is a good example of the limitations of the sea. Overly rich waters (river pollutants) led to overpopulation of plants and animal life, followed by oxygen shortages, massive algae dying, suffocated fish, and the threat of poisoned shellfish.

Need to Inventory the Seas

This kind of report would lead us to the conclusion that while the potential yield is definitely limited, it is at the same time remarkably vague. In his book *The Hungry Planet,* George Borgstrom, professor of food science at Michigan State University, calls for resolving the vagueness. "Before the next step is taken —the exploitation of the oceans through organized sea culture —a cooperation of global dimensions in development work is mandatory, with a research effort of dimensions hitherto undreamed of. . . ."

He went on: "The number of research vessels needs to be multiplied many times over. Submarines would have to be rebuilt for use in systematic subaquatic investigations. The Soviet, U.S. and Japanese deep-diving . . . bathyscaphs are required in much larger numbers. Nobody can justly claim that the world cannot afford this."

He concluded by saying that the world needed an *oceanwide intelligence-gathering system "in order to draw from the seas the massive amount of information, data, and experiences required to put together a coherent picture of the sea"* (italics added).

First Significant Ocean Images

Almost a decade later, most of the underwater tools which he had visualized still have not come into being. But even as Borgstrom wrote in 1965, the first steps were, almost inadvertently, being taken at Cape Kennedy to develop the other aspect of his suggestion: an ocean intelligence-gathering system.

At an interim briefing a few days before the launching of Gemini 4, astronaut Ed White was handed a list of points on the earth for which various scientists had requested photographs. In those days, geologists were still the most enthusiastic devotees of space photography (Gemini 4 would be the mission on which White would take the now-famous picture of sand dunes in North Africa). By contrast, there were but few requests on White's list for ocean pictures, although there was one view of the Florida Keys proposed.

We have seen how space photography in those days had a low priority at the Cape. But Ed White, even though on the brink of becoming the first American to walk in space, nevertheless found time to do his homework for his minor picture-taking mission. (When he returned from Gemini 4, his film expended, he brought back the portfolio of pictures which so impressed Dick Underwood.)

Among the few photographic words of advice given to the early astronauts was the suggestion to shoot the ocean only when the sun's rays were oblique to the earth, in order to avoid the reflected glitter from the water. When Gemini 4 arrived over the Keys, however, it was high local noon, and the glare directly up from the water was inescapable.

However, since there was nothing better to shoot at the moment, White went ahead and clicked off three exposures with his Hasselblad, being careful not to overlap the areas covered by any of them. As with his African desert picture, White's tendency to keep snapping the shutter again was fortunate. When Dr. Paul Lowman later looked at the second of the three pictures, he noticed that the sun's glitter—squarely in the center of the field of view—brought out the movement of the ocean in a striking way. Lowman's carefully educated eye could detect two separate and distinct patterns in the waves, some of them straight lines, others *circular*.

Lowman showed the picture to an oceanographer, Dr. Robert E. Stevenson, then of the U. S. Bureau of Commercial Fisheries, who was immediately intrigued by what he saw. Stevenson first recognized the straight lines as waves moving routinely along with the wind. The circular pattern, on the other hand, had been formed by water moving against the wind, whipped into high waves topped with just a thin cream of foam. It was a rougher sea, and it reflected the excessive glitter quite differently from the straight waves. Stevenson studied the circular patterns and identified them as counterclockwise eddies. The eddies were peeling off the Florida Current as it surged through the Straits of Florida, curling their way back through the Keys.

The Secret Migration

"What a fantastic picture!" Stevenson reacted zealously, for reasons not immediately obvious. As he explained it to me later, "The eddies were not the only thing it showed us. I was able to see a set of extremely long waves, say, three miles long, bending around two points in the water. I pulled out a chart and found that two sea mountains were charted directly under the points where the waves were bending. According to the chart, the mountaintops were six hundred feet below the water surface, and yet they caused a bending in the waves that we could see in a picture taken a hundred miles up!"

Stevenson concluded that the high-glitter areas, although admittedly poor places for peering deep into the water, seemed to be ideal for revealing the motion of water going against the wind. "At this point I began to believe that satellite pictures could teach us some things about currents, eddies, divergences and convergences—the whole field of ocean dynamics."

After seeing that picture, Stevenson immediately began to review all Gemini photography and related the pictures to his own fisheries work. Oceanographers had long been puzzled as to how shrimp eggs, or larvae, arrived each spring in Florida Bay. Every summer and fall the famous Florida pink shrimp were caught in large numbers in their spawning grounds north of the Dry Tortugas, hundreds of miles west of the bay. Then the following spring the next generation of shrimp would mysteriously appear in Florida Bay. The bay was their nursery, where they grew to adulthood before crawling back across the ocean floor, westward with the current, in their dutiful migration to the Tortugas. This adult migration everyone understood. But the question had always been: How do the shrimp eggs, not widely known for either great intellect or swimming prowess, later manage to buck the east-to-west current and reverse the trail of their parents? How could they float eastward from the spawning grounds to their nursery in Florida Bay?

Ed White's picture had seemed to offer the answer. While the main current flowed westward along the Keys, the eddies were curling back to the northeast, threading back through the Keys into the bay. These, Stevenson had been convinced, were the routes that carried the eggs to the nursery.

He had arranged to sample the waters between the Keys and check the samples for shrimp larvae. "It was like counting the whiskers in a beard," Stevenson said. "The water was loaded with larvae."

This was the reason why Stevenson had been so ecstatic when he had first seen the picture of the eddies. As he had suspected, the migratory route of the eggs had been outlined in White's picture, and it might now become a gold mine in studying the pink shrimp. Once the migration was fully understood, ways would be found to forecast the shrimp crop and consequently lead to its better utilization.

The Surging Gulf Waters

At this point he returned to his office in Galveston, Texas, where the Bureau of Commercial Fisheries scientists were busy forecasting shrimp availability each season. The area of importance was the adjacent Gulf of Mexico, with its complex coastal waters and its fertile shrimp grounds.

Stevenson's next opportunity for satellite photography was in 1966 on the last of all Gemini flights, Gemini 12. This time he was granted comprehensive coverage of the Gulf. The flight took place in November, and whereas White's picture had coincided with the larval migration, Gemini 12 would capture the last phase of the shrimp cycle. The adult shrimp had completed their spawning and were now in the process of getting themselves caught and hauled into shrimp boats. This November picture would give the fisheries bureau an opportunity to relate the Gulf currents to the habits of the adult shrimp.

In a turnabout from earlier briefings, Gemini 12 astronaut Buzz Aldrin was *encouraged* to shoot high-glitter areas, and some of them were preselected and plotted for him by Stevenson. As it turned out, Aldrin succeeded in bringing back the preselected photos and more, all of them found to be excellent.

Quite apart from the glitter pattern, Aldrin was also advised to take coastal shots which might reveal sediment movement. The rivers' outpourings onto the shallow coastal shelf surged along in the shallow water, like a pack of terriers, darting in all directions at once. The pictures Aldrin took captured the brown lines of river sediment, conveniently dyed by Nature and frozen by the camera into precise outlines in the Gulf. They revealed patterns in the coastal waters which oceanographers had been trying to unravel for decades.

Locating Good Shrimping Areas

Checking past catch records, Stevenson determined that the eddies, which moved generally seaward, were the good shrimping areas. They were apparently chosen by adult shrimp feeling the compelling urge to crawl out into the ocean and spawn and seeking the most favorable current to get there.

Stevenson thereupon approached the Texas Shrimp Association with his charts of the Gulf and told the shrimpers he thought the eddies were the factor creating their high-catch areas. They listened with more interest than surprise, Stevenson recalls, with one of them remarking, "That sounds right. I always go where the water is muddiest." Shrimpers were generally close-mouthed about their haunts, but the speaker turned to another shrimper and asked, "Don't you?"

The response was an immediate nod. "I always have, but I've never known why before." This casual agreement convinced Stevenson that satellites were telling him the truth.

For the next two years the shrimpers and the fisheries bureau exchanged information, the shrimpers reporting their catch locations, bureau scientists plotting them. "Their best catches continued to be along the paths of the eddies located by Gemini 12," Stevenson remembers.

The next satellite picture of the area was not until Apollo 7, two years later, again in the fall of the year, and again the shrimping season. Since the good shrimp catches had continued to occur in approximately the same areas, no one was surprised to find that the axis of each of the eddies had moved only about five miles since the Gemini 12 picture. "At that point," said Stevenson, "I was convinced we had a tool that could detect the current on the water surface and forecast the actions of sea life on the ocean floor hundreds of feet below."

As a scientist, Stevenson was convinced that he had developed a practical tool, but I wondered if the shrimpers really regarded it as such. To find out, in 1973 I chatted with John Mehos of the Gulf-based Liberty Fish and Oyster Company. Mehos' company had been active in the work with the fisheries bureau in the 1960s. The information the bureau had supplied had been accurate, he remembered, but he added, "In a sense, though, it was no big thing, because we pretty well knew the good shrimping areas of the Gulf long before satellites came along.

"However," he pointed out, "if we had been shrimping in some virgin area, it would have been another story." Oceanographers know that almost every continent of the world has abundant shrimp beds which have never been precisely located—in the Persian Gulf, in Australia's Gulf of Carpentaria, off the coast of Africa's Guinea or South America's Guiana. These are areas where a single satellite picture might now open up new food sources.

Seasonal Shrimping Forecasts

By 1969, Stevenson and other confident oceanographers were now looking forward to Apollo 9, which was to be a March orbit, and the first opportunity to photograph the Gulf during the season of egg migration.

They were not disappointed in what followed. On the day of the Apollo picture, the calm winter current seemed to be influenced more than they had expected by the north wind. Coordinating the picture with observations in the water, Stevenson theorized that the March current was a direct product of wind direction; during days of southerly winds, a south current would carry the larval shrimp into the bays and estuaries, their nurseries, where they were to develop as juvenile shrimp. On days of north wind, however, the current would be reversed to ebb seaward, and those larvae, unable to battle their way upcurrent, would prematurely terminate their careers as shrimp.

Stevenson checked the catch records and the wind data for the past ten years and found close correlation between the number of days of southerly winds and the shrimp crop for the following autumn. This relationship should make it possible, said Stevenson, by noting the number of days of south wind in a larval season, to predict closely the following season's shrimp crop, six months away!

It apparently worked. This was another matter on which I consulted John Mehos, and this time he said the satellite information had been truly practical. "Ever since we got that information," he said, "shrimping forecasts have for the first time been accurate enough to be taken seriously. We use them in planning our season—buying materials, alerting the union to probable hirings, and deciding how to distribute our boats for the year.

"The canneries and processors use the information in much the same way," he went on. "The size of the upcoming crop tells them how many shrimp to hold over, and whether to contract for fruit and vegetables to fill up the slack in a light shrimp year." If a satellite never sensed the area again, the formula to improve the shrimp forecast each year had been developed, and it would remain fairly constant, even though the dynamic ocean might change to some degree.

The Changing Face of the Sea

And change it would. "Inward and outward to northward and southward, the beach lines linger and curl . . ." Sidney Lanier had written a century ago about the constantly varying coastal waters. This dynamic quality of the seas gave oceanographers a special interest in satellite imagery; the most recent maps of much of the world's land and water were fifty years old. "That's great for geologists," Stevenson said. "They can take a picture today and then relax till the next Ice Age shakes things up. But the ocean won't let us wait that long. It's *always* on the move."

To learn the movement of the ocean, as man must if he is to harvest its wealth and survive within it, he must never stop watching it. Oceanographers have always known this, but until satellites came along, they didn't know how to do it. The oceanographer needs to forecast tides and currents and ocean storms on a worldwide basis.

For him to do this, a model will have to be developed and fed regularly with satellite data.

An annual survey of ocean movement will be necessary even in the middle of the oceans where changes come slowly. In more turbulent ocean areas and in coastal waters, monthly or even daily changes need to be noted. And in sometimes violent waters like the Gulf Stream, fluctuations should be monitored several times daily. Oceanographers visualize a continuous satellite monitoring system with input from visual, IR, and, eventually, microwaves. Offering the only means of penetrating cloud cover, the microwave sensor would be an extremely useful member for any oceanographic-oriented satellite. Several concepts for ocean satellites have been considered at various times, but the most viable current proposal is SEASAT, suggested by NASA (for the late 1970s) in hearings for the 1975 budget. SEASAT would use microwaves and laser to measure wave height, tides, and currents, and would climax a long series of ocean experiments which have been occurring ever since Ed White's Gemini 4 pictures.

During the years since Gemini 4, satellites have risen from obscurity in oceanic research to a leading role. We should interject, however, that at least a handful of oceanographers foresaw the importance of remote sensing by aircraft even back in the mid-1960s; one of these is Dr. Gifford C. Ewing of the Woods Hole Oceanographic Institution in Massachusetts. Ewing for years has picturesquely likened the seas of the world to a thin sheet of letter paper. He points out that the ocean's depth, like the thickness of the paper, is roughly only 1/6,000 as great as its surface area. And "like the sheet of paper," he comments, "much of [the ocean's] information is written on its face and exposed to view from afar."[2]

Temperature and Chlorophyll

A phenomenon which Ewing particularly expects to detect on the "face" of the ocean is photosynthesis, in which the sun and the sea conceive plant life. This face, or photic layer, the key to all ocean life, is, rather conveniently, the very same portion of the sea which remote sensors are capable of penetrating; of the top hundred feet or so of the ocean which supports photosynthesis, as much as fifty feet can sometimes be observed by sensors! Ewing long ago felt that man could perhaps detect the photosynthesis by sensing both the *temperature* and the *color* on the face of the ocean.

Temperature is important because cold water from deep within the ocean is frequently lifted to the surface in an "upwelling," and along with the cool water come nutrients. Once this fertilizer reaches the sunlit upper ocean, Nature puts photosynthesis into motion and grows her garden of sea vegetables. And where the vegetation blooms, the fish will not be far away.

As the garden grows, the sea becomes slightly green, indicating the possible birth of chlorophyll. Color, then, if it could be detected, would seem to be a logical signal of photosynthesis; if a temperature sensor were already detecting a cool upwelling of water and now a second sensor indicated a color change in the same area, we would have a ringside seat to watch the garden grow.

Man through the ages has detected striking color differences in the Red, Yellow, and Black seas, and Ewing maintained that sensors could detect variations many times more subtle than the human eye, and use such variations to measure chlorophyll.

Locating High-Probability Fishing Waters

In the summer of 1967, Ewing took off on the first of several temperature- and color-sensing expeditions, studying the difference of water as he flew southeast from Woods Hole. With a plane equipped with optical sensors, he flew over Nantucket Sound, across the nutrient-rich fishing areas on the coastal shelf, then over the cooler, sterile Gulf Stream. An oceanographic vessel chugged beneath him.

Both the ship and the aircraft measured temperature and color in the fertile fishing banks and in the less-productive Gulf Stream. Measuring the color in blue, green, and red bands, and correlating it with temperature, the expedition brought back a wealth of data. Blue and green bands, as well as temperature, appeared promising. *When evaluated, they offered strong indication that aircraft were capable of locating gardens of phytoplankton and seaweed.*

Then in 1972, two NASA engineers, John Arvesen and John Millard, expanded Ewing's experiments in an expedition that took them over half the world, from the California Current to the rich fishing areas off northwestern Africa. Flying in the *Galileo* (Convair 990), they maintained altitudes ranging from 1,000 to 35,000 feet. Building on Ewing's work, they recorded temperature and concentrated their color sensing on comparisons of green and blue bands.

Instead of having to analyze their data after the trip, as Ewing had done, they operated in "real time," and were able to read their sensors at the same moment they flew over a point.

They noticed that fishing boats appeared beneath them only when their gear was showing high algae probability. This casual relationship between temperature and color readings, on the one hand, and fishing boats, on the other, was hardly a correlation to excite a statistician. But the two engineers were convinced they were on the right track in finding fish.

They were also enthusiastic about their real-time capacity. Their ability to be "hot on an event" would be useful in fishing, because the life of an ocean garden can be extremely brief. A typical upwelling of nutrients may occur on Monday, to be followed by a blooming of algae on Friday, and be completely consumed by fish and zooplankton the following week. What the NASA engineers were perhaps developing was *a method to predict high-probability fishing areas, several days in advance, over rather wide areas.*

Forecasting Fish Species?

Since certain sea plants prefer warmth, and others coolness, some biologists visualize a day when chlorophyll and temperature readings will be able to point to an individual species of vegetation. And if we could distinguish a particular vegetation, we would be just one step away from identifying the chain of marine life that would feed on it. Color and temperature would indicate when a garden was sprouting, and a particular fish would be expected to follow; obviously a nice arrangement for managed fishing.

Before achieving any such high-powered selectivity, however, there are still problems in the elementary sensing of chlorophyll. The scattering of the reflected light limits color sensing to a low altitude art—difficult from higher airplane altitudes and still out of the question from a satellite. But even if limited to aircraft, chlorophyll might eventually prove to be the highest-priority message on the face of the sea.

Whereas chlorophyll sensing has not yet made the jump to satellites, temperature sensing has. Two of NOAA's latest weather satellites, NOAA-2 and NOAA-3, have since mid-1974 been doing an effective job, in limited areas, of collecting the surface temperature of the ocean at intervals *as close as half a mile.* While up until now a fisherman has always faced hundreds of miles of sea ahead with no water temperature reported, NOAA-2 and NOAA-3 now offer the potential of showing him where each degree of variation occurs.

This new temperature information is what albacore fishermen have dreamed about for years. Albacore, very temperature-sensitive fish, will dance along the 62-degree side of a temperature gradient rather than swim over the "edge," as the fishermen call it, into the 63-degree water. For that reason, if the fishermen can locate the edge of the 62-degree water, they are likely to fill up their boats with albacore on the spot.

The albacore industry has not yet taken advantage of the new temperature data, but, in one of the ironies of bureaucracy, the Inter-American Tuna Commission—working in behalf of tropical tuna fishermen, who really don't care much about temperatures—has. The commission, in its La Jolla, California, station, has for years been transmitting valuable weather charts to the tunamen, generally scattered throughout the eastern Pacific, via a radio facsimile system. Incorporated in the radio "facs" weather charts have been random ocean temperatures, which were collected from all the ships at sea long before satellites became available. In between these scattered reports from ships were hundreds of open miles where the meteorologists in La Jolla had to guess ("interpolate," as weathermen prefer to say) at temperatures in order to complete their charts. This interpolating has been more than accurate enough for the tunamen, who have never paid much attention to temperatures anyway. "The *weather* information on the 'facs' we like," one tunaman in San Diego told me. "To know about storms is very nice. But water temperatures—who needs them? We find tuna in any water above 70 degrees, and it's always warmer than that south of here where we fish. We don't need a satellite to tell us the water is warm in the tropics."

Unabashed by this disinterest among the users, the theoreticians at the tuna commission are now assimilating the intensive NOAA-2 temperature information (for limited areas in the Pacific) with the idea of eventually including it on the "facs." Instead of having an oceanful of estimated temperatures to ignore, tunamen can soon ignore the results of the world's most elaborate temperature-gathering system.

In fairness, we should say that despite the present lack of use for the system, some of the tunamen will probably experiment with the information. "Perhaps we should be paying more attention to temperature," says August Felando, manager of the Tunaboat Owners' Association. "We may learn something."

Dr. Merritt Stevenson (no relation to Robert) of the tuna commission agrees. "The boundaries between two water types are important in the ocean. Where two water fronts converge, you find a collection of flotsam (logs and debris) and, quite likely, fish." Stevenson thinks that albacoremen have become aware of this fish-congregating tendency, because they fish close to shore, where water fronts are smaller and front edges therefore much more apparent. Now that tropical tunamen can be kept informed of where the edges are, they may begin experimenting and conceivably find that a valuable new technique exists for finding tuna.

Dr. Robert Stevenson looks on temperature as important, but for a slightly different reason. He believes that temperature, by pointing out nutrient areas, could soon become an operational tool, not so much for locating where fish actually are congregated today, but for determining fishing areas weeks or months in advance. Government and commercial organizations are becoming interested in finding data on *this* month's ocean plant life, as a means of forecasting next month's fish havens. Satellites, Stevenson believes, are not far from finding the phytoplankton gardens much as Ewing and Arveson located them with aircraft—but with temperature, rather than color, as the key to the hunt.

NOAA-2 and NOAA-3 theoretically can monitor all the ocean temperatures of the globe, but a vast computer data bank will be required before that system can ever become a reality. A more immediate method, however, may have been uncovered by Skylab in 1973 and 1974, when Robert Stevenson and other Skylab investigators made a breakthrough in studying the sea face. They discovered that wave patterns of the sea could serve as indicators of water temperature. The connection between wave patterns and temperature became apparent in Skylab photos when Stevenson noted cumulus clouds scattered in broad circular patterns over the Caribbean. Their patterns were apparently created by the temperature variations of the ocean below. Stevenson looked further at the high resolution Skylab photos and noticed outlines of waves in the water itself. The waves clearly indicated eddies following the same patterns as the clouds above them. The eddies were caused by no apparent disruption of any physical nature and seemed to point to the conclusion that they, like the clouds, were related to varying temperatures in the ocean surface.

To investigate this possibility, U. S. Navy aircraft in January 1974 were assigned to fly over the areas where the eddies had been sighted, and were timed to coincide with Skylab 3 passes. The aircraft dropped rows of small bathythermographs (capable of recording temperatures in the top thousand feet of the ocean) along the paths of Skylab tracks. The bathythermographs indicated that the surface water of the eddies was several degrees cooler than that of the surrounding sea face. The eddies appeared to be upwellings from layers of cool water three or four hundred feet below, where nutrients were concentrated; the cooler layer, in this case, was acting as a nutrient trap.

Robert Stevenson considered the eddy-temperature relationship to be at least as significant as his observations of the eddies in the Gemini pictures nine years before. This time, however, the eddies were apparently not caused by wind, nor by islands or other disruptive quirks of geography; they seemingly were signatures of cyclonic disturbances in the sea which created temporary upwellings of nutrients, identifiable by the wave and cloud patterns.

If such eddies can be observed on a worldwide basis, they offer the potential of predicting fishing zones as they globally appear and reappear. Unlike the promising but complex NOAA system of charting temperatures, this mapping of eddies may offer a simple means of locating fertile upwellings on a broad basis. "By tracking nutrient areas, such a system could reduce hunting time for some species of fish by fifty per cent or more," Stevenson calculates.

Stevenson further believes that such nutrient areas may appear in unexpected places; various long-reputed "sterile" waters, such as the Gulf Stream, might prove to have sporadic fishing caches never before suspected. The temperature areas may also prove to be surrogate signatures for spawning areas. So, whether tunamen are yet excited about satellite sensing or not, oceanographers are bound and determined to prove their case.

Turbidity as a Message on the Sea Face

While temperature would presently seem to be a more advanced tool than color, insofar as satellite sensing is concerned, there is one particular aspect of satellite "color" sensing in which progress has already been made: the distinguishing of dark water from clear water. We have previously seen how streaks of sediment revealed currents in Buzz Aldrin's Gemini pictures of the Gulf coastal waters. However, some attention has since been paid to the importance of the sediment itself, as it relates to fishing. Suspended bits of sediment and living organisms tend to reveal their presence by muddying the ocean water in which they float, and this darkened water then provides a haven where fish can both eat and hide. Lower order fish instinctively attempt to break the food chain by keeping out of sight to avoid being eaten.

Menhaden, a low-order fish which has become popular recently among chicken-feed manufacturers, has long been suspected of hiding out (from its predator, the bluefish) in the muddier waters in the Gulf of Mexico. In an effort to evaluate this theory, an ERTS experiment was conducted in the Gulf in 1972 by the Earth Satellite Corporation and the Marine Fisheries Service. Five one-day studies were conducted, examining ERTS pictures of the Gulf and comparing them with the day's catch of menhaden. Fishermen cooperated by reporting and locating their catch on those test days. The results, if not amazing, were conclusive: Higher fish catches were recorded in areas where sediment was high, indicating a correlation between the fish and the muddy waters.

While the experiment only confirmed what fishermen already knew, it proved something else that was much more significant concerning our ability to read images. Interpreters could readily separate the dark areas from the surrounding waters in satellite pictures; *even in waters as shallow as two fathoms* (twelve feet), where the dark ocean bottom caused interference by showing up distinctly, *interpreters could distinguish the turbid water from the clear.* The project's ERTS investigator, Dr. Paul Maughan of Earth Satellite Corporation, was extremely pleased with the results. "During the seventies, man should be able to develop turbidity as a key to large concentrations of fish. And since fish populations move so fast, only a satellite can coordinate such an oceanwide picture," he stated.

Fish Oil and Other Aircraft Techniques

Wave motion, color, temperature, and turbidity thus all seem to be fairly reliable indices for studying the ocean surface. Together they provide a good set of criteria for cross-referencing, and if science were to develop only this set of tools, remote sensing should be able to go a long way. But oceanographers seem to be bitten with a virus of restlessness and are constantly looking at the face of the sea with an eye for new indicators of life below. One of these indicators is a product of the fish themselves: fish oil. An oily film given off by fish becomes noticeable on ocean surfaces and has been successfully sensed from aircraft. The greatest obstacles thus far seem to lie not in sensing the oil, but in deciding what it means.

Followers of the fish-oil concept have serious ambitions about being able to develop signatures for various fish species and to provide instant recognition of the haunts of anchovy, tuna, albacore, etc. Such sophistication will require greatly improved interpretation before real effectiveness can be achieved.

However, as more and more such unique criteria achieve even marginal effectiveness, their combined value increases considerably faster. Examining an ocean area with several techniques would allow a technician to go from one criterion to another, much as an organist changes keyboards. If color, for example, fails to yield a reading, perhaps a combination of temperature, turbidity, and even fish oil will lead to some conclusion.

The discussion of these techniques inevitably leads to the story that satellites will soon be serving, as seaplanes do now, in scouting for fish, tracking them, and dramatically radioing advice to the fishing fleet. Newspapers and congressmen have contributed to the myth by talking about "satellites that can see fish in the ocean." A delightful fish story (but hopelessly untrue). It is reminiscent of the (very true) counting of raisin trays in Chapter 8, described by the surveyor as being "like counting pickles in a barrel." However, at 570 miles over the ocean—as opposed to 17,000 feet over Fresno—we can't manage to see a pickle barrel, let alone single pickles; there simply is no satellite-sensing program aimed at spotting single fish, or even schools of fish. What satellites can do is detect high-probability waters in which fishermen are likely to discover "pickles" by the barrelful, i.e., congregations of fish.

To be sure, there do exist ways to sense the actual fish themselves—but from platforms considerably closer to the water than satellites. Fishing companies in the Gulf of Mexico have made progress in detecting fish at nighttime, from boats and even low-flying planes, by searching for bioluminescent organisms. Certain ocean areas possess microorganisms which, when agitated, illuminate themselves in much the same manner as glowworms. Often vast numbers of them surround a fish as it slashes through the water. When agitated, the organisms turn on and light up the fish just as light bulbs illuminate the outside walls of skyscrapers. A low-light-level television camera can be trained on a school of fish and sense them if they are swimming within a few feet of the water surface. A sensor on a boat, viewing a school just fifteen or twenty feet away, can actually distinguish an individual fish. And an aircraft sensor operating on a dark night at altitudes under 5,000 feet can distinguish a school as a single blob of light. But to equip a satellite several hundred miles away with the enormous optics needed to pick up such luminescence would seem to be an expensive way to run a fishing fleet.

Another popularized idea is the "beeper." Transducers, rather like the beepers attached to cars in television spy stories, are now available and could theoretically be used. A device the size of a two-cell flashlight can be attached to a large fish. The beeper can be quite easily interrogated as to its position whenever the fish breaks water, with less effective response when the fish uncooperatively remains submerged. While these devices have been used with some success on whales, which are above water much of the time, they have yet to prove effective on fish.

The concept of the beeper is to release a single, radio-equipped fish and track it by satellite in the ocean where hopefully it will join up with large schools of its unsuspecting fellows. These craftily conceived devices, which tend to irk more sporting-oriented individuals, show more promise as a research tool than as a fishing device. Researchers, sporting or not, need all the techniques available if they are to observe what is occurring in the sea.

Food Sources Not Yet Exploited

While we recognize the food capacity of the ocean as a very finite thing, we also suspect that there are occasional harvest bonanzas not yet being pursued. The tremendous anchovy fishing grounds in the Humboldt Current off Peru is an example of a fishing grounds developed almost overnight. Completely unknown just a few years ago, that part of the Humboldt is virtually alive with anchovy and is the reason why Peru now lands more fish† than any other nation in the world. Created by the cool, mineral-upwelling Humboldt, the waters are an example of the kind of area which temperature-sensing satellites will be discovering or defining in the future.

Another example of a temperature phenomenon originates in waters a few hundred miles to the north of the Humboldt, where U.S. fishermen operate. Just once every few years the area is blessed with the warm, wintertime El Niño Current, in which various fish thrive. El Niño originates in the Pacific off Central America and affects fishing for hundreds of miles; a constant monitoring of her waters would make fishermen aware of great potential catches which would otherwise go completely wasted, because while some waters of the world are being steadily overfished, others are underworked.

Some waters are, in fact, believed to be bulging with varieties of sea life never harvested commercially. A small Antarctic shrimp, *Euphausia superba,* is suspected of being the most abundant unexploited delicacy in the ocean world. Various biologists, including Frank Alverson of Living Marine Resources, Inc., in San Diego, California, have estimated that more than *50 million metric tons* of this species could be harvested each year. If true, this one shellfish could virtually double the world's entire annual seafood crop, which today stands at 62 million tons!

Alverson made this prediction while we were standing along the southern California coastline. As an afterthought, he added, "You don't really have to go to Antarctica, though." He nodded southward down the beach. "A few miles down there"—he pointed—"off Baja California, the UN estimates that a million tons of red crabs are going to waste every year."

† Peru leads in numbers of (small) fish landings, although not in total fish tonnage.

The value of satellite information for exploring the remote Antarctic is of course obvious. But Alverson would have us believe that even in such well established fishing haunts as Baja California, the satellite's contribution of preliminary information, to be followed more intensely by other means of sensing, would have great potential.

Conservation and Management

Hand in hand with the exploiting of sea life must be the conservation of it. Any managed water must be continuously inventoried to keep alive the chain of life, to forecast population changes quickly, and to regulate activity there.

For years, various maritime nations have joined together in international commissions controlling the catch of tuna, salmon, halibut, and others. Fishermen have thereby become accustomed to sensible regulations limiting their season, their catch, and their areas of operation—all with an eye to conserving fish for future years. Satellite information can soon be a part of all this. If, for example, a satellite system were to forecast a prolonged temperature drop in a managed ocean area, then tightened regulations might be indicated for the following year's catch.

Managing ocean areas, developing and controlling certain fish within geographic limits, is a growing concept which should eventually lead to full-scale fish farming, with satellite monitoring of managed areas part of the technology that will get us to that point. However, there may well be intermediate phases. As soon as satellite information becomes usable enough to have true commercial value, a national or international fish forecast center could be established for the exclusive use of boats and fleets which cooperate. Boats could be limited in their visits to managed areas where species of seafood were being developed or where conservation techniques were being practiced. Advice regarding controlled harvests of various fish would be given only to those fishing fleets complying with regulations. Centers could maintain surveillance of all the cooperating fishing boats in the world by using a foolproof transducer (beeper) system—which performs considerably better on the hulls of boats than it does on the bodies of fish. Those boats which failed to remain on-limits would be blacklisted and not receive subsequent ocean data. This system could work even without full cooperation from all governments. There need be no penalties or confrontations on the high seas. The system, internationally administered, could be simple: No beep, no data.

Most, but not all, maritime nations seem to recognize the need for conservation. Man's ignorance of just *what* he should be doing to conserve his fishing grounds is, more often than not, the biggest problem.

A generation ago, the California sardine was an important factor in the Pacific marine life. A fishing fleet and several companies existed solely to harvest and market this one rather valuable fish. Then, in just a few seasons, the sardine disappeared, its ecological niche eventually occupied by the less valuable northern anchovy. Oceanographers were unable to determine whether its disappearance was caused by a temporary break in the ocean life cycle or was simply a matter of overfishing and underregulation.

Today, what may be a similar crisis, on a local basis, faces the Dungeness crab off San Francisco. In order to save this extremely valuable species, man must understand the environment in the crabbing areas. While California fish and game authorities have not sponsored an ERTS or a Skylab investigation, they are nevertheless poring over all available satellite data for possible clues to the crab's disappearance.

These and innumerable problems like them will continue to face us, sometimes as natural evolution, but more often as a result of changes man has made in Nature. As satellite data banks develop profiles of normal ocean conditions, we should become able to recognize exceptions to the norms as soon as they arise.

Sensors Opening Door to Ocean Study

If the late 1970s do indeed produce oceanographic satellites, there will be a backlog of work awaiting them. They will be required to deliver data to the National Oceanic and Atmospheric Administration (NOAA), using most of the sensing techniques mentioned in this chapter. By then it will have been well over a decade since Georg Borgstrom made his appeal for "systematic subaquatic investigations" and "bathyscaphs in much larger numbers." Yet bathyscaphs will still not be in the water, although hopefully some will be in the offing. However, satellites could draw from the seas "the massive amount of . . . data" Borgstrom said was necessary for a coherent picture of the sea. Remote sensing—a tool which Borgstrom did not anticipate—will have laid the immense groundwork at a tiny fraction of the cost of any method using the immense underwater trappings he suggested. It will have selected the areas where more intensive underwater study, of the type he visualized, can be concentrated in the future. Oceanology will have finally begun. Satellites will have taken the first major step toward developing the harvest of the sea.

And perhaps most important of all, for our future food planning, we will have commenced to learn both the potentialities and the limitations of the oceans.

11. The Weather—"Acts of God"

The earth, when viewed from thousands of miles out in space, is outlined in a sky that is clear but for the thin layer of weather next to the earth itself. The world is enveloped by this ten-mile layer of troposphere where winds turbulently soar about, sucking up millions of tons of water from the oceans, carrying it thousands of miles, condensing it, eventually dumping it as generally useful precipitation, and, in the process, zapping man's fragile world with incredible amounts of damage.

If a chef were to make a giant meatball and somehow wrap a tight layer of spaghetti around it, he would have a physical model of the earth and its bizarre air currents. A dynamic global swirl of wind swishes hot and cold air, water vapor, and pollutants around in wild spaghettilike patterns which man had simply never seen until the satellites of the 1960s.

It is a picture capable of bewildering most of us, and it may well have terrified some weathermen in the 1960s, showing them for the first time what they had to contend with. But most of them took it in reasonably good humor and, public opinion to the contrary, have even utilized it to upgrade forecast accuracy by a few percentage points. Fundamentally, however, the weather guessing game is the same. Denver's snowstorm today still is the forecast for Boston's day after tomorrow.

However, the National Environmental Satellite Service (NESS, a branch of NOAA) insists that better things are on the way. Not too many years from now, the weatherman will be anticipating that Boston storm when an updraft of water vapor occurs at the equator the week before. After more than a decade of experimenting with satellites, with models and computers, today's meteorologist is convinced he can pull these tools together and make sense out of the spaghetti. At long last, he now has his first operational weather satellites, and he fully expects that within a few years he will be as skillful at forecasting a week or more ahead as he now is for two days—at least where normal air fronts and storms are concerned.

The same meteorologist, however, is much less certain about improving his forecast of today's hurricane in the Gulf of Mexico, which could jump all over the Southeast and then become a flood in Virginia just three days hence. And he faces an even tougher forecasting job in the Oklahoma tornado, which is bred and born in an afternoon and lives for only an explosive quarter of an hour. As more than one weatherman has said, the meteorologist who can predict exotic storms will have the world's most challenging forecast in the palm of his hand.

The Twister

Since 1970, Ed Ferguson has become well acquainted with the National Severe Storms Forecast Center in Kansas City. The center has the job of keeping weather forecasters throughout the country aware of major storms breaking, usually in the area the forecasters call "Tornado Alley," the central and southern states. Radar had been their biggest tool in plotting the storms, but in 1970 NESS asked the center if it would consider using satellite pictures as a backup system.

The National Environmental Satellite Service is located in Suitland, Maryland, where it is in the business of doing what its name implies—obtaining pictures from satellites and providing them as a service to various governmental weather organizations. The storms center in Kansas City is a typical "customer" for NESS, whose service would not only include the transmitting of satellite pictures to Kansas City—but also the lending of Suitland personnel to the storms center to interpret the pictures for the regular forecasters and to try to make believers out of them. Among the personnel assigned to go from Suitland to Kansas City was Ferguson, a young but experienced satellite meteorologist, a Missouri native with the twang still in his speech. (His selection was probably no accident. If you want to sell a picture in Rome, send a Roman.)

The pictures Ferguson was to interpret were the same kind of satellite imagery which has since become well known to viewers of commercial television: the black glossy prints with white clouds over a white outlined map of the United States. The pictures come from the Applications Technology Satellites (ATS series), the weather spacebirds which hover 23,000 *miles* above the earth. That extreme altitude is chosen for the satellite in order to increase the orbit substantially. At 23,000 miles, the immense orbit requires twenty-four hours, thereby swinging the satellite around the globe at precisely the speed the earth itself rotates. Hence it has no motion whatsoever in relation to the earth, but seems to hang suspended over a given point of ground. It is thus synchronized with the earth, or "geosynchronous"—as motionless as an albatross viewing a ship from the mast. ATS-1 is geosynchronized over the equator in the Pacific, due south of Christmas Island, while ATS-3 hangs over the equator of South America, its circle of vision including all of North and South America except for northern Canada and Alaska.

The ATS duo were actually launched as communications satellites (ATS-3 being a replacement for an aborted ATS-2), with a spin-scan "camera" sent piggyback on each ATS as a modest weather experiment. Every twenty-five to thirty minutes during daylight hours, the scanner is trained on the hemisphere and the picture it takes is transmitted back to earth. When the first transmissions took place, the astonishingly good results caught everyone unaware. "We were so surprised it was embarrassing," said NESS meteorologist Lester Hubert. Nevertheless, NESS quickly overcame its embarrassment, geared up to handle the pictures, and promoted their use among NOAA weather organizations.

Although the first pictures were amazing, considering the distance from which they were taken, they were not yet in the same class with the radar images which weathermen were accustomed to using. When he was assigned to Kansas City in early 1970, Ferguson had realized the limitations of those first ATS pictures, and he must have inwardly winced as he showed them to Allen Pearson, director of the storms center. Pearson, Ferguson realized, might be moderately impressed by the feat of taking pictures from 23,000 miles, but he would be much more concerned with picture quality. And, quite frankly, the pictures were fuzzy, as Pearson was bound to notice.

Pearson looked at the first pictures with interest. "They're fuzzy," he pronounced.

The message that then went back to Suitland was loud and clear. The customer wasn't buying. Things had to be improved. Ferguson and NESS had two jobs to do: upgrade the pictures and learn to interpret them effectively. Intensive efforts were made in both directions.

Several weeks later, on July 24, 1970, an unusual opportunity presented itself in Kansas City. On that day, a key radar failed to operate, making the satellite the principal tool for the forecaster. In a morning satellite picture, a narrow squall line of clouds appeared over Nebraska and South Dakota. It was four hours before any radar detection was available, and for those four hours forecasters cautiously utilized the satellite picture. They plotted a consistent flow of satellite information on their charts and found it quite usable. Eyebrows were raised speculatively.

By late afternoon, everything was breaking loose; the sky was in turmoil. The results: twelve tornadoes, fifteen hailstorms, and several forecasters who were cautiously enthusiastic about the information from ATS.

That incident triggered a turn to the weather satellite, not as a device to replace radar, but to complement it, especially in the earliest detection of storms. Traditionally, tornado "watches" have informed residents in the Midwest and South whenever tornado-producing weather was in the air. Unfortunately, as every resident of that Tornado Alley knows, the announcements—which always alert an area hundreds of miles wide—have no more effect than a boy calling "wolf." For several months each year, forecasters have always inundated these states with watches, which all but a few people have accepted with a "ho-hum." Hence when a twister actually *has* ripped through one square mile of Kansas (and avoided the rest of the state), the few people caught by the tornado have been unprepared, while most of the rest of the state has become further immune to watches.

Pearson decided the usefulness of satellite pictures revolved around their ability to detect the tornadoes as they were being formed, before radar was able to detect them, allowing the forecaster to release a definite warning for a small area, instead of a general watch for an entire state. If the satellite pictures proved sharp enough, they could perhaps offer urgent *localized* warnings that the blasé folk in Tornado Alley would take seriously.

Picture quality continued to improve, and by 1971 forecasters were able to plot tornadoes in isolated fifty-mile-wide warning boxes, which generally proved accurate. Pearson then decided ATS-3 was a member of his forecasting team.

Then came March 28, 1972, one of those gray Missouri days that Ed Ferguson remembered from his youth. The temperature outside the seventeenth-floor Kansas City weather office was moderate. But the forecasters today were more interested in the skies over Texas than those over Missouri; a warm lower air mass was trapped in Texas, with cold upper air and a certain amount of moisture in the lower atmosphere completing the formula; the forecasters were talking Texas tornadoes.

Ferguson had in his hand a satellite picture which had been transmitted from ATS-3 just minutes before. The spin-scan camera in ATS-3 had made its image of most of North and South America by scanning the hemisphere, a line at a time, simultaneously transmitting each line to earth. The image was received on Wallops Island, Virginia, and instantaneously retransmitted to Suitland and on to Kansas City in real time. The negative which the photo specialist in Kansas City routinely pulled off his machine thus actually showed the way the earth had looked from space only seconds before. The only delay in the system occurred in Kansas City, where humans stepped into the process and used fifteen minutes to make a black-and-white print from the negative.

The print in Ferguson's hand was, like the pictures used by TV stations, only the U.S. portion of the total image. The photo specialist had labeled it 1427 Greenwich time (8:27 A.M. Central Standard Time in Texas).

Ferguson looked at the central and eastern areas of Texas which the forecasters had been watching since early morning. Texas was generally clear and black, except for a white wedge of clouds piercing it from the south. The wedge was a jet stream of moisture that would further trap the warm air already being squeezed into the sky over Texas. The eastern Texas area was ripening for a broad-area warning. Over the next five hours it was to receive two; "wolf" was thus cried twice, with the usual lack of effect on the populace, but that was all the warning that was possible at the moment.

Then, at 1:30 P.M., Ferguson was handed his eleventh half-hourly ATS picture of the day. It looked much like the previous ten, but his eye immediately want to eastern Texas, where the center had issued the warnings. He saw a small wispish line of clouds extending north and south through the warning area.

Another satellite meteorologist, Bob Fett, was visiting Kansas City that day and was standing beside Ferguson, half hoping to see some twisters. Fett saw the line, too, but shook his head. "That's not our action," he predicted.

"Agreed. But I think it's just down the street from it." Ferguson saw enough to know that the wisplike squall line was a harbinger of tornadoes. He informed the forecaster, and radar was alerted.

While Ferguson and Fett were looking at the glossy print, the photo specialist was making further use of the satellite negative. As each satellite picture had arrived during the day, he had reduced it to 16-millimeter size and spliced it together with the earlier pictures; this gave him a mini-motion-picture film that showed the day's weather progression in just a few frames of elapsed-time photography. He was now adding the 1:30 P.M. picture. The mini-film was looped into a circle, so that it could run continuously. A few minutes later, Ferguson and the forecasters were watching the motion picture as it repeated itself again and again, showing, in just a few seconds of film time, exactly how the clouds had formed and moved since dawn. The film was showing them the weather as it had developed throughout the United States, although their interest remained mainly directed to eastern Texas. Of the entire map, it was the area of the Texas squall line that kept attracting their trained eyes, causing them to search for something that wasn't yet there.

Then at 2:51 P.M. a new satellite picture showed something different. The squall line was still there, but so was another small line of clouds some sixty miles farther west. To a casual observer, the second line would have appeared to have been a carbon copy of the first. But the forecasters could detect subtle differences—clouds were rounded out from the second line, like a string of beads, and one of the beads was supporting a tiny cumulo-nimbus anvil top, streaking off to the east. Through the eyes of ATS-3, Ferguson and Fett were seeing a tornado, stretching itself, getting ready to break out of its cocoon.

They didn't wait for the cocoon to break. *Within minutes of the picture, a sixty-mile-wide area in Angelina County was being given severe tornado warnings. Residents took cover.*

Within the next hour, several small communities in Angelina County were torn apart as swirling masses of air moved through their streets, carrying along roofs, bathtubs, and telephone poles. The record shows that twelve houses and nine mobile homes were completely destroyed. The destruction all came in the box the Severe Storms Center had anticipated. As a result of preparedness, no lives were lost.

Bob Fett had watched the satellite's participation in the

day's tornado forecasting with satisfaction. The forecasters, he noted, had been particularly sold on the satellite's contribution. "We had them eating out of our hands," Fett commented later.

The storms center continued to feed—and thrive—on satellite information, and two years later the satellite/radar team immortalized itself in forecasting methodology. On April 3, 1974, the most widespread explosion of tornadoes in U.S. weather history ripped in all directions along a line from Canada to the Gulf coast. Storms center forecasters found themselves advising local weather stations over eleven states. Had such an onslaught occurred in the pre-satellite era, extensive refined warnings of the tornados would never have been possible. As it was, however, ATS pictures were used to coach radars onto the fast-developing storm lines, and warning zones of some fifty by ten miles were established in advance of tornado after tornado. Residents showed no hesitation in taking cover. Pearson said, "Three hundred twenty-eight lives were lost, but a thousand more may have been saved."

The one fault Pearson found with the satellite information on April 3 was that it was available from ATS only in daytime. But when a new satellite with IR nighttime ability, the Synchronous Meteorological Satellite (SMS-1)* was launched six weeks later, Pearson was enthusiastic. On May 17, 1974, just a few hours after the launch, he said, "*We* got it up today," displaying an apparent feeling of sponsorship for the geosynchronous type of satellite he had regarded rather doubtfully in years past. The GOES series, he now believed, was bound to effect another tremendous advance for tornado forecasting in the South, where nighttime storms were especially common. The tornado warning technique which Bob Fett had watched Ed Ferguson developing in 1972 had proved itself time and again in daytime, and was now becoming an around-the-clock operation.

* The May 17 prototype was named SMS-1 by NASA, but the subsequent series for NOAA would be called Geosynchronous Observational Environmental Satellite (GOES), the name we shall use hereafter.

The Hurricane

Fett had been an early product of the space age, and he, even more than Ferguson, had watched it grow from the stage of random "gee whiz" satellite pictures to today's serious role, although in the mid-1960s there had been times when he had become very pessimistic about weather satellites.

On April 1, 1960, the first experimental weather satellite, Tiros I, had been launched. Just nine days later, it had aroused the meteorological world with the photograph of a typhoon† near New Zealand. "I was an air force weather pilot going after my master's degree in meteorology," Fett remembers, "and that one picture excited me enough so that I built my entire thesis around it."

As a veteran weather pilot, or "hurricane hunter," graduate student Fett quickly visualized satellites as something that could revolutionize searches for hurricanes. Satellites would, he believed, make it unnecessary for hurricane hunters to drone across the skies in needle-in-the-haystack searches, looking for storms which seldom occurred. The tedious patrols of straight-and-level flight ("boring holes in the sky," as pilots called it) could be a thing of the past, he thought. Satellite pictures now could handle the preliminary search, coaching pilots into areas where they could spend their time more productively, inspecting known hurricanes. To Bob Fett it sounded as if a "dream combination" of satellites and aircraft was in the making.

He would seem to have been thinking out loud; at that point the Air Force apparently decided Fett should see his dream in action, and he was sent to Guam as a weather pilot, where he could see satellite pictures in actual use.

Fett was in for shock in Guam. The pictures were clear

† Called typhoons in the western Pacific, hurricanes in the Western Hemisphere, and cyclones in the Indian Ocean, along with numerous local names.

enough,‡ but to his dismay, he found them inaccurately charted. "This is really something," he told a debriefing officer after a fruitless mission. "You send me on a vector where I'm supposed to find a typhoon and there's nothing but blue sky. And meanwhile, here"—he was pointing at the status board—"we find a typhoon churning away like a locomotive. More than 120 miles from where you and that satellite sent me!" The weather people in the 1960s were learning what other users of earth observations would face several years later in the ERTS era: remote sensing was not magic, but was a tool to be meticulously developed.

And developed it gradually was. The hurricane hunters did, in time, find that satellites indeed were taking the place of hole boring. Within a few years, air force and navy hurricane hunters actually *were* spending their flight time inspecting storms already located by ATS-1 and ATS-3. Such an incident occurred in 1969. ATS-3 pictures had enabled weathermen at Miami's National Hurricane Center to track Hurricane Camille in the Caribbean long before pilots detected her and in fact for a week before she even reached hurricane status. The Miami hurricane center, which has storm responsibility over-water similar to the Kansas City over-land assignment, was able to continuously track hurricane Camille and estimate her intentions. Pictures continuously flowed in from ATS throughout the daylight hours, and hurricane hunters kept a closer watch by air. A computer digested the input concerning Camille's course and speed, and then spit back predictions as to where and when she would hit the shoreline. Although the predictions were crude, compared with the modeling expected in the late 1970s, it was most useful. As a result of the predictions, some 50,000 persons were evacuated to safety from coastal areas —areas where Camille, a few hours later, was to take 200 lives, sink or ground nearly 120 vessels, and destroy hundreds of millions of dollars in property.

But although the warnings of Hurricane Camille probably

‡ Resolution that might appear "fuzzy" in looking for a one-mile-wide tornado is "clear" for a 300-mile-wide hurricane.

saved at least hundreds of lives, everyone agreed that satellites were not exercising their full potential. The lack of ability to take pictures at night was frustrating; much of Camille's 170-mph furor was developed the night before she struck the shore, and neither aircraft nor satellite observed her course and speed during these critical hours.

Fortunately, nighttime pictures became a reality in mid-1974, when the first prototype for the new Geosynchronous Operational Environmental Satellite (GOES) series was launched. Weathermen were ecstatic over the detail they found in the first infrared pictures; and the nocturnal coverage which appeals to Allen Pearson for his tornado hunts will be just as popular for tracking hurricanes.

Another shortcoming of ATS pictures has been that although they offered a live, real-time image of a hurricane, weathermen have been able to interpret very little about her. They have been able to see where she was at a given moment and where she had already been—but anything else has had to be shaky conjecture. Research to develop forecasting models has been badly needed.

Some of this research is now coming, from studies of the satellite imagery itself. We have already seen how geosynchronous satellite pictures are useful partly because of their real-time quality, which can give a meteorologist a look at the world of that very moment (if he doesn't mind getting his fingers wet on soggy prints). However, the pictures have a value long after they have dried, in post-storm research for techniques that may lead to better forecasting. Enough storms have had their pictures taken now so that archives are bulging with documentation on how hurricanes develop. We know of half a dozen meteorologists doing research with film loops and still pictures.

One student of the pictures is Suitland's Andy Timchalk. After studying the movements and course changes of numerous hurricanes on film, Timchalk has developed a theory to predict where a hurricane is likely to go next. By measuring the movement of clouds on the periphery of hurricanes, Timchalk determines wind speed and directions of gusts. By plotting these wind vectors geometrically, he attempts to predict a hurricane's travel plans for the next day. His formula is still being evaluated.

While Timchalk is concerned with storm direction, another Suitland meteorologist, Vernon Dvorak, has studied cloud pictures with the idea of predicting a storm's development (or lack of it). Dvorak's system calls for a dozen or so measurements and observations of a storm's size, shape, and quality—all to be funneled into a formula to predict the hurricane's growth. Dvorak's formula is likely to become operational soon.

Pictures are not the only form of satellite data used after the fact to analyze hurricanes. The experimental Nimbus satellites at one time carried IR sensors capable of sensing water vapor. In the 1960s a wealth of vapor data was tucked away in government files and was recently "discovered" by researchers. Observations of the classic Camille showed a great lack of vapor within her a few hours before she intensified dramatically from 100 to 170 mph. Does a hurricane inhale vapor just before rapid intensification? The answer to that question could be vital to future hurricane watchers wondering if a storm is preparing to fall apart or move ahead full bore. (The water-vapor data theory is not confined to hurricane study. Meteorologists have found that water-vapor sensors can sometimes detect jet streams at particular altitudes in *clear air*. This could mean an increase in opportunities for routing airliners both to take advantage of tailwinds and to avoid the extreme turbulence which turns some passengers green.)

A more controversial theory is being explored by two scientists at the Naval Postgraduate School in Monterey, California. Meteorologist Dr. R. J. Renard and oceanographer Dr. Dale Leipper have asked for satellite help in studying ocean temperatures—an attempt to predict hurricane movements. It is well known that a hurricane requires a water temperature of 79° F or greater if she is to stay alive. She will quickly dissipate over cooler water or land, the warm ocean water being the aphrodisiac that continually triggers her.

Renard and Leipper suspect that a hurricane not only needs

but *seeks out* the warm stimulant, and they reason that this may be why hurricanes often take wild unexplained turns. They had hoped to test their theory with a water-temperature survey from Skylab sometime during the South Pacific or Gulf hurricane seasons of 1973. If Skylab were overflying a hurricane, astronauts would be asked to use the MSS to check water temperature on all sides of the storm, so that for the first time scientists would be able to see if a hurricane followed a warm path in preference to other water. It would be a dramatic first for science—looking over a hurricane's shoulder and detecting the water temperatures around her.

Unfortunately for research, the 1973 tropical-storm season was extraordinarily calm, at least during the weeks that Skylab was aloft. Ironically, when the astronauts did at last encounter one hurricane (Ava), it was at a time that Skylab's MSS was malfunctioning.

Leipper says they now anticipate pursuing the experiment with ocean temperatures recorded by NOAA-2. If results are revealing, more comprehensive experiments could follow throughout the late 1970s. If the theory proves valid, it will provide a great leap forward in hurricane forecasting.

Wind—The Key to Forecasting

While ocean water can perhaps be a factor in hurricane forecasting, it is wind that is the key to all weather prediction. If we could comprehensively monitor wind movement around the globe, we could expect to forecast the location of storms and air masses with accuracy unknown today. One project which will probably be carried out by GOES in the mid-1970s concerns a system being developed at NESS by Les Hubert for *determining winds at varying altitudes*. Hubert has had much to do with the elapsed-time film loops which we saw in use at the Kansas City storms center. These film loops, showing changes in the sky every half hour, are of course capable of indicating the direction of cloud movement. By projecting the loops on a viewing table, the movement of clouds in successive frames can be plotted so that wind speed and direction can be calculated in all those areas covered by ATS-1 and ATS-3 pictures.

Unfortunately, however, the loops tell us nothing about the altitude of the clouds, and we have no way of knowing if the wind we are calculating is at 500 feet or 50,000 feet. The height of the clouds is obviously a critical third dimension for understanding where and how weather fronts will collide or interrelate.

Hubert's project is aimed at using GOES to sense the IR rays emitted by clouds, to determine cloud temperatures. The next step, based on the fact that cooler clouds are formed at higher levels, is to translate cloud temperature into *approximate* cloud altitude, on a worldwide basis. For such complex data, the meteorologist's charts and pictures are inadequate; he requires mathematical formulas of the weather before such involved processes can be conducted. Weather, like all other phenomena of nature, can and must be expressed and predicted in computerized mathematical models.

We must keep in mind that a scientist is usually a unique individual. He seemingly would prefer to describe a spring breeze or the growth of a rosebud in mathematics rather than in poetic French, or in English, or even in scientific German. "Mathematics is really the language of nature" is the way Arctic scientist Bill Campbell puts it, confidently adding, "The Great Designer was obviously a mathematician!"

The idea of models in weather forecasting is not new. It was born in World War I when an ambulance-driving English Quaker and mathematician, Lewis Richardson, spent his time at the front, between battles, observing the weather. Trying to determine the interaction of many such factors as motion, water, and heating and cooling rates, he developed the first formulas for weather forecasting. Today, his lengthy equations are regarded as classic.

Unfortunately Richardson himself was never to see his equations successfully utilized, for two reasons. First, he lacked the technology to make calculations swiftly enough to be used. Once he had gathered the data, it took him months to solve his equations manually; obviously a handicap in daily weather forecasting. Since the late 1940s, of course, computers have been overcoming this problem by offering an increasing degree of calculating speed, and computer models are on their way to a leading weather role.

Richardson's second difficulty was a lack of accurate weather observations. Once again, time has improved things; virtually every country in the world has become involved in the attempt to acquire weather information, with thousands of ships, land stations, and balloons providing data constantly via telegraph. Yet in spite of this monumental cooperative effort, three fourths of the globe is still virtually untapped for information. And ironically, of the hundreds of balloons released every day, virtually none is released in the Pacific; the Atlantic is the area which has been the subject of almost all international weather treaties.

"Every schoolboy knows our weather goes from west to east," one of the weather service's leading meteorologists on the West Coast of the United States repeatedly grumbles. "But where are all our weather stations? In the Atlantic." The Pacific has few islands, and only one weather ship exists there, midway between Hawaii and the mainland United States. Furthermore, the cost of installing and maintaining adequate stations for this immense ocean could quickly creep into the billions. Fortunately, a much more thorough job can be done at considerably less cost by a satellite capable of covering the entire globe—a *polar* satellite.

The Weather Team—Geosynchronous and Polar Satellites

While TV viewers see pictures nightly from the geosynchronous satellites, most viewers have never heard a word about the weatherman's polar satellites. Yet polar satellites have been in existence longer and in fact have been operational since October 1972. At that time, the first successful operational weather satellite, named ITOS-D, was launched into polar orbit by NASA. As soon as the satellite was found to be electronically alive and well, NASA allowed her sister agency, NOAA, to adopt it. The NOAA weathermen modestly renamed the satellite after themselves, calling it NOAA-2, engendering some confusion in the process. It succeeded NOAA-1, an earlier, short-lived satellite, as well as numerous experimental polar satellites. (The most recent experimental satellites were the ESSA series, ESSA being the parent agency for the weather service until NOAA was formed. Other predecessors of NOAA-2 have been the several Nimbus satellites, a still-current experimental series.) The important thing now, though, is that the NOAA series constitutes the weatherman's first operational satellites. NOAA-3, as we noted in an earlier chapter, has arrived as a backup for NOAA-2, and all future operational *polar* satellites will presumably be called NOAA-something (at least until the agency next changes its name).

With the first operational geosynchronous satellites (GOES) replacing the experimental ATS, the operational weather team is complete. Polar and geosynchronous satellites will satisfy all the concepts of the weatherman for as far ahead as anyone can see.

NOAA-2 is in a near-polar orbit similar to that of ERTS and Nimbus, with an altitude somewhat higher than ERTS (900 miles compared to 570). Like ERTS, it orbits north to south on the sun side of the world, south to north on the dark side. It takes some 114 minutes to make a complete orbit from north to south pole and back again, during which time the earth rotates almost 29 degrees to the east. Its tracks over the earth are thus 29 degrees apart, and in the course of twenty-fours hours, the same point on the earth's surface has revolved back under NOAA-2's orbit. Whereas ERTS senses only a 100-mile strip, NOAA-2 senses the entire 29-degree sweep (almost 2,000 miles). Thus, while ERTS is designed to spend eighteen days between observations of a given point, NOAA-2 covers every point each day and again each night. It has considerably less resolution than ERTS, a tradeoff for the wide sweep and the advantage of being able to observe the entire world's weather twice a day.

Soundings and Accurate Wind Altitudes

The most significant assignment for NOAA-2 is in the field of *soundings*, an expanded form of remote sensing. Whereas sensing is generally concerned only with studying points on the surface of the earth, soundings examine the column of air that rests upon each of these earth points. When NOAA-2 takes a sounding, it is sensing the air above a given earth point up to a height of some ten miles (that is, to the top of the earth's tropospheric layer, wherein all weather lies).

By sounding the ten-mile-high column in various IR wave lengths, scientists are learning to determine precisely the temperature not only at the earth's surface *but at any given point in the column.* By then feeding these temperatures into models, meteorologists can also determine pressure for every point in the column. And since pressure maps are a time-honored tool for making approximate wind predictions, wind maps can be created for the various altitudes. "The sounding is thus a profile of temperatures for the column. Since each temperature is accurately fixed as to altitude, the end result is an approximate forecast of winds at precise altitudes for the next forty-eight hours," Suitland meteorologist Harold Woolf explained it to me.

That sounded like a duplication of Les Hubert's work, and I said so.

"Not really. Just the opposite, in fact," Woolf answered. "Hubert's cloud pictures are supposed to supply an approximate altitude for his clouds and an accurate wind." And NOAA-2, I had just learned, was to provide approximate wind and accurate altitude.

Was there any way that we could combine the two, I wondered, and get the best of both worlds?

"We expect to. We find that temperatures are, rather handily, accurate in *both* the sounding and the cloud pictures. So we match up the temperatures from both systems. When Les Hubert's clouds offer us an accurate wind, we pick out a nearby sounding. Somewhere on that column of air we find the temperature corresponding to the cloud temperature. That's our altitude."

"GOES offers the wind and NOAA-2 furnishes its altitude, then?"

"Right. We should be able to do this for any spot on the globe." This kind of accuracy on a worldwide basis is precisely what weathermen since Richardson have dreamed about: global wind information at all altitudes. With it, weathermen will have worldwide modeling capable of precisely forecasting weather movement; without it, they will have more of the guesswork of the 1960s.

The nations of the world have historically cooperated in international weather efforts, most recently in the Global Atmospheric Research Program (GARP). A multitude of non-space-oriented efforts in European countries already show signs of making GARP a success. But satellites must be the critical building blocks if GARP is to dramatically expand the world's present forty-eight-hour skill in forecasting.

The human and economic savings in improved weather forecasting are so gigantic in scope that no unanimity exists in estimates made to date, although all projections jump quickly into the billions. A National Academy of Science study considers only the four most visible areas—agriculture, construction, transportation, and flood control—for the United States alone and estimates that long-range weather forecasts would save $2.5 billion annually. A much more comprehensive (seventeen-volume) study made by IBM Corporation claims that if weather forecasts could be 95 per cent correct even three days in advance, man could save $60 billion a year![1]

The Do-It-Yourself Picture Kit

The IBM study assumes not only that the weather forecasts are accurate but that man responds; unfortunately this is not always true. A tragic example of the lack of response to a satellite warning occurred in Bangladesh, then East Pakistan, in November 1970, when 300,000 persons were killed by a typhoon and flooding. Lack of ability to mobilize evacuation equipment for persons stranded in rice fields and on thousands of marshy islands rendered the two-day advance notice provided by ESSA-8 useless. (Being outside the geosynchronous satellite coverage, Pakistan depended almost entirely on ESSA-8, the polar satellite operating at that time.)

The satellite warning, which under different circumstances might have prevented that terrible tragedy, was sent *directly* from ESSA-8 to Dacca, Pakistan, by means of a unique system called Automatic Picture Taking (APT). Some 500 of these APT units have been distributed throughout the world. This is a kind of do-it-yourself satellite-picture-receiver kit, costing less than $10,000 each, including antenna, receiver, and auxiliary equipment. Every twelve hours, the current polar satellite (now, of course, the NOAA series rather than ESSA-8) passes over any given area and transmits a vidicon weather picture of the locale, which APTs can receive and put to whatever use is desired. These low-cost units are not restricted to weather bureaus or even to government bodies. Any individual with a yen to receive satellite pictures could set up shop with an APT as easily as he could become a ham radio operator.

Many of the 500 APT systems in use are in underdeveloped countries such as Pakistan. However, some of the units which have proved most effective to date have been in the United States; one such system is located at the Inter-American Tuna Commission station at La Jolla, California. The tuna commission meteorologist regularly receives and processes NOAA-2 pictures of the local area, which, in his case, is the eastern Pacific.

As mentioned in an earlier chapter, a weather chart is then prepared and radioed in facsimile each afternoon to the tuna ships at sea.

Most of the Pacific tunamen in the United States are Portuguese Americans, from families that have fished, or whaled, as far back as family history can be traced. In the Pacific, tropical tuna is the prize which entices them to sea, and they fish from Catalina to the Galápagos, docking their boats in San Diego. One of them is Ed Silva.

Silva had grown up fishing in the Atlantic, out of Massachusetts, on his father's boat. He became one of the pioneers in the San Diego fleet, first putting his boat in the water there in 1926. Now he spends most of his time working at the tuna-boat association by the pier. One spring day in 1972, I asked Ed if he had found the Pacific a calm place after the Atlantic.

"Well, no," he answered. "Not when the *chubascos* are moving."

"*Chubascos?*" I was puzzled until Silva explained that *chubascos* was the name given to Pacific hurricanes by the Mexicans. It was fair enough to let the Mexicans name them, I decided, since they usually bore the brunt of them; if a *chubasco* didn't hit Mexico, it usually never reached land at all. In fact, the U.S. weatherman had often gone unaware of them in the pre-satellite days. But satellites had started keeping track of them, and found that there were often more hurricanes in the eastern Pacific than in the Gulf of Mexico. These hurricanes have an average diameter of three hundred miles, and in 1971 nineteen were charted. "The most I can remember in one year," noted Silva.

"How often do tuna boats get caught in these *chubascos?*" I wondered.

"Before we had radio 'facs' we usually lost five to six boats a year," Silva said. "But now, *nobody* gets caught. Twenty-five boats get the 'facs,' and they warn the rest of the fleet when a *chubasco* is on the move."

The unfortunate natives in the rice paddies of Bangladesh

would never be able to dodge storms the way fishermen could. But for people in a position to react, the worldwide APT service was an amazingly low-cost and effective warning device.

Dramatic Information from Polar Regions

APT, with all of its popularity and potential, was an afterthought for polar satellites. The real forte of the polar satellite is, not surprisingly, polar observation.

In sensing the north and south poles, the meteorologist sees two vast worlds of ice, snow, and water, which, to him, are a pair of giant weather generators. The weatherman has learned that the amount of ice at the poles varies considerably from year to year. In winters such as 1971–72, when the tongue of ice extending down along Newfoundland was greater than anyone could remember, scientists variously estimated the winter's ice volume at 10 to 40 per cent above an average ice year; the broad span of those estimates is an indication of how limited our ice knowledge is. Weathermen agree on such unstartling bits of logic as a belief that "years of cold air and years of heavy ice seem to coincide." There is, however, considerably less agreement as to which causes which!

Weathermen generally agree that during the dark polar winter, the long IR rays emitted by the water provide more than a thousand times the meager radiation given off by the surrounding ice. So in years of less ice, considerably more heat (and also more moisture) escapes from the sea into the atmosphere. By the same token, years of heavy ice causes drier, colder air to develop. All this becomes dramatically relevant to the United States when the air rolls down from the north pole to collide with the warmer Canadian air mass and explode into storms all over North America.

The effect of the ice is great, but as of now the facts leave room for varying theories and questions. A fairly prevalent belief is that in years of heavy ice, IR emissions from the reduced water area would be lessened to the extent that the world would experience a slight temperature drop. Also popular is the concept that a heavy *reduction* in ice might start a chain reaction in warmer weather (that is, less ice leads to more IR emissions and vice versa), melting the icecaps of Antarctica and Greenland, and raising the sea level.

Some physicists and meteorologists embrace a more complex theory, which we will not develop here except to summarize its result: An absence of ice would ultimately create meridional air currents flowing from poles to temperate areas and, strangely, creating a tendency for more extreme weather, *both hot and cold,* in the temperate zones. The same theory leads to a conclusion that the meridional currents would cause frequent cyclones on a global basis. The physicists who hold such views fear that these phenomena might be triggered by polar pollution and ice melts.

We cannot prove or disprove these theories so long as our vision is limited to the temperate zones; most of the answers to these intriguing questions lie hidden in the icebound vaults of the Arctic and Antarctic. An explosion of knowledge useful in short-, medium-, and very long-range forecasts is sure to take place once we have unlocked these two ends of the world.

And suddenly it seems possible to do this. With NOAA-2 and its far-IR ability, we can penetrate the polar umbrella and see the Arctic in the wintertime. In spite of the continuous winter darkness, IR sensors can distinguish between the cold ice on the one hand and the sea water with its outpouring of hoarded IR heat on the other. Forecasters will note the amount of ice and snow at the poles and observe its effects. After a comparison of ice volume is made for a few winters, weathermen will devise models to reveal how cold the air will be, where the moisture will develop, and how it will move.

By 1980, enough experience should have been gained so that the ice information should be yielding good short-term storm forecasts. And seasonal forecasts will have become a serious venture for the first time ever.

Unlike soundings, which are offering the weatherman a well-defined data source, the character of ice information is still vague at this point—but the impact of the new data is not. Soundings and polar information will become the right and left hands of the weatherman. Together, they will move us toward the GARP goal to extend today's *forty-eight-hour forecast accuracy up to two weeks by the early 1980s.* Weathermen, rarely heralded by society as among the world's great prophets, hope to be coming into their heyday, thanks largely to polar satellites.

GOES Emerging as Flood Tool

Long-range forecasts, no matter how outstanding, will always need buttressing in regard to local phenomena. Once again we call on the geosynchronous satellites. The ever-recurring problem of spring floods will surely be attacked with increasing data from GOES.

Each year dozens of U.S. rivers pose the threat of wiping out huge segments of the populace along their shores. In such years as 1965, flood losses of hundreds of millions of dollars are reached, with swollen rivers sprawling out across midwestern cities and farms.

In 1965, the first of several stages (we'll call it Phase One) of evolution in flood-fighting techniques occurred. Volunteer citizen workers were organized as never before to build earthen dams and form sandbag brigades. Frequently their efforts were effective, but other times they came too late. I remember taking part in a typical situation in Rock Island, Illinois, one Sunday afternoon that year, when the Mississippi River was nearing its crest. Townspeople were ardently trying to secure the Rock Island Mill Street levee, when we ran out of bags and gave up. Most of us retreated to high-land homes and, in the American tradition, watched the flood on TV as we awaited more supplies. While we waited, the Mississippi did not; by evening we were watching Chet Huntley showing a video clip of the flood overrunning our dike.

Four years later, Phase Two of flood fighting was introduced. ATS snow pictures and other kinds of weather warnings alerted the Army Corps of Engineers, early in the winter, to the likelihood of 1969 surpassing all previous records and becoming one of the flood years of the century. The Corps responded by building dams and organizing volunteers with a foresight and planning never known before. Sandbags were available, not the afternoon after they were needed, but weeks *ahead* of time. Weeks later, the wild-running waters arose as predicted, but in most cases had trouble jumping the riverbanks in populated areas. So while 100 million dollars of property was swept away, another 250 million was saved.

It was progress but still not an ultimate solution. The 100-million-dollar loss could have been further reduced by even more accurate and earlier data. The precipitation gauges and snow-depth recorders which kept the engineers informed generally had to be manually read and were a weak point in the system.

Phase Three. By 1972, when several states were deluged by flash rain floods, wireless precipitation gauges equipped with transducers were tied in to local warning stations. The results in many cases were truly impressive; lives were saved because alarms were sounded and evacuations performed.

Nevertheless, much better control of the emergency would have existed if readings could have been available from more remote areas and if many more gauges had existed. The Corps of Engineers had a question. Why couldn't a geosynchronous satellite monitor all the gauges and transmit the readings to the river forecast officials in *real time?* The satellite could operate as a continuous communications link between the flood areas and the River Forecast Centers.

Phase Four. With these questions in mind, an experiment is being coordinated by the Corps of Engineers, using ERTS as the communications link. Twenty-five data-collection platforms in New England waterways accumulate data on stream flow, precipitation, temperature, soil moisture, tide changes, snow depth, and water quality—and transmit it twice daily to ERTS (much as volcano data is collected). The readings go from ERTS to Greenbelt, Maryland, and, by landline, to the Corps of Engineers computer at Waltham, Massachusetts, arriving in real time. Early reaction from the Corps to the experiment has been that satellites could provide information not only much more conveniently than all-land gauge systems—but cheaper as well. Engineers expect to get the flood and other river information from satellites for a few thousand dollars, whereas hundreds of thousands would be needed to cover New England effectively from the land.

The coordinating of thousands *or even millions* of gauges by a satellite is not as difficult as it may sound, rotating as it can through a series of thousands of electronic interrogations in the time a human observer would need to record one guage reading on a clipboard. Such a rapid satellite monitoring would have been invaluable in the 1973 Mississippi River flood, when repeated rainfall deluged and redeluged areas from March to May, making it difficult for the Corps of Engineers to know where to concentrate its efforts next.

Polar Satellites and Floods

However, Phase Four might not ever need to come into being. In the field of flood control, satellites as communications links may be surpassed by satellites as remote sensors: Phase Five. Instead of radioing measurements from scattered points, sensors might sometime measure *all* the moisture in an area. IR sensors already offer a certain ability to measure snow and ice for water equivalency and to signal when snow and ice are melting.

Unfortunately, however, IR doesn't have the ability to penetrate heavy clouds to measure the floodwaters below (and, notably, flash floods seldom occur under clear skies). We have indicated in previous chapters that IR sensors, the darlings of the mid-1970s, might be particularly eclipsed before 1980 by another generation of technology, the microwave sensors. Not only can microwaves work effectively through clouds, but they do some of the same things as IR, and do them much better. Physicists at the Aerojet Corporation say their microwave equipment has the potential to measure moisture accurately—on snow-covered fields, in a lake or stream, or in an icecap. They say they can even penetrate the top level of the soil and measure the saturation of the ground. (Microwaves would thus offer a highly sophisticated version of Dr. Mark Meier's ERTS project discussed in Chapter 5.) If this microwave system can be effectively developed for satellites, it could be one of the most useful techniques in remote sensing, leapfrogging other water-measuring systems on the drawing board. After all, who would want to interrogate a system of gauges to feed into a model if one could read the answer from a microwave sensor in one fell swoop? The microwave method would quote the total water volume present in an area—along with the amount of runoff expected after the thaw. *It could provide a constant report on the water available to descend on a particular river, dam, or community.*

Numerous microwave moisture-measuring experiments in aircraft have already been conducted. Now satellite data from Skylab and Nimbus 5, and eventually Nimbus 6 (with a late-1974 launch), is to be utilized to develop the technique further.

Whether microwaves are in the immediate offing or not, the young field of space science appears to be working mightily to help the not-so-young science of hydrology find its way out of the dark ages. But the observing of floods, much as the study of such other "acts of God" as hurricanes and tornadoes, has just barely begun, and meteorologists are not anxious to conjecture about how fast progress will come. On the other hand, more conventional weather fronts (comprising some 98 per cent of our storms) are, as we noted earlier, factors in which the weatherman truly expects to improve his prognostications. With the cloud picture from GOES, and with the soundings and the ice and snow studies from the NOAA satellite series, he expects to add many days to his routine forecasts by the 1980s.

21. Acts of Man

The eight million residents of the Los Angeles basin are, on Memorial Day, a strong second to the racers of the Indianapolis 500 in their camaraderie with the not-so-open road, gasoline engines, and exhaust fumes. The remainder of the year, the Angelenos rank as undisputed champions.

With this characteristic presumably well in mind, the director of the Environmental Protection Agency (EPA) stood before television cameras in the middle of Los Angeles, on January 15, 1973, and suggested that, as a means of reducing smog, gasoline and automobiles should be virtually eliminated. With all the confidence of a Greek slave asking a crowd of angry Romans to spare a fallen gladiator, William P. Ruckelshaus told the Angelenos that an 80 *per cent reduction* in gas usage was his formula for meeting the new Clean Air Law. (It would be months before rationing would be discussed nationally relative to a completely different problem, the energy shortage, and the concept of virtually throwing away their car keys sounded ludicrous to the natives.) Whatever his intentions in making the proposal, in which he was said to be less than completely serious, the EPA director kept a straight face throughout his interview.

Another character in what the press considered a clean-air charade was Dr. James N. Pitts, Jr., director of the Air Pollution Research Center at the University of California at Riverside. Within a few hours of Ruckelshaus' statement, Pitts told a television audience he welcomed the idea of national debate on gas rationing. Tongue in cheek or not, Pitts said he saw rationing as a means of reaching fundamental decisions—"decisions that will affect not only southern California but the entire nation, since many of the major urban areas face serious problems of photochemical smog." (Pitts's research center was statewide in scope, but Riverside, located fifty miles downwind of Los Angeles, had been one of the cities which had gone to court to require EPA to clean up the city of Angels.)

Underlining the national impact of the smog problem, the Stanford Research Institute on the same day had released a three-year study, describing agricultural smog damage; it indicated that the greatest crop damage occurred not in long-time smog pacesetter California, but in industrialized eastern states.

In spite of national overtones of the smog problem, however, the worst air pollution in the East was factory-inspired, different from California's automobile-developed photochemical smog. If gas rationing would be effective in controlling smog anywhere, it would be here. Ruckelshaus had made his proposal in the right place, although the Los Angeles reaction was hardly an endorsement.

"Worse than prohibition," said A. J. Haagen-Smit, chairman of the State Air Resources Board.

"Like throwing the baby out with the bathwater," added Frederick Llewellyn, president of the Los Angeles Chamber of Commerce.

Whatever the psychology for their respective stands, Ruckelshaus and Pitts retreated discreetly, Ruckelshaus beating a hasty exit to Washington and Pitts retiring to the friendly surroundings of Riverside.

Man Hesitant to Control Weather Overtly

The fact that the Angelenos would not readily acquiesce to the order was predictable, although perhaps not much more so than it would have been for the people of the rest of the United States. Americans as a whole are probably not much inclined toward laying down one's driver's license to alter the composition of the air (an energy shortage possesses at least somewhat more clout). Weather and environment are responsibilities man still defers to higher authority, rather than taking them on himself. This is ironic since earthlings have inadvertently been affecting air quality for decades, indeed for centuries in some places, by spewing vast tonnages of effluents into the atmosphere. But now to follow a deliberate policy of altering air or weather with premeditation is somehow a very different concept to man.

Even those actively engaged in smog reduction doubted that such a massive experiment as an 80 per cent cut in gasoline was a reasonable approach. The idea of virtually capsizing an entire megalopolis, putting businesses and workers out of touch with one another and causing an estimated 400,000 unemployed—in one bold, experimental stroke—was not considered a logical move by most of those concerned. Dr. Pitts, for one, had a better idea (which he elected not to introduce during the EPA charade); his research center was then a year and a half along with a four-year NASA program providing airborne sensing equipment for smog study.

Immediate Sensing

Twice a month a twin-engine NASA plane equipped with immediate- rather than remote-sensing equipment, flies the Los Angeles basin and measures the air immediately around the aircraft. Following an established saw-tooth pattern, it drops to near ground level at prescribed check points and climbs to 10,000 feet at others. During flight, its equipment measures carbon monoxide, nitrogen oxide, and other air constituents with which it comes in contact. (None of these constituents has ever before been extensively monitored at varying altitudes.) In addition to the in-flight sensing, chemists also make what they like to call "discreet grabs" during the flight, filling containers with air and analyzing it later for hydrocarbons.

The over-all objective is to develop a Los Angeles basin smog model, in which various control strategies will be simulated. "By seeing what happens to the air under different conditions, at various altitudes, a meaningful model can be created and we can begin to define the problem," explains NASA research scientist Hermie Gloria. Instead of attempting to manage flesh-and-blood southern Californians (a formidable challenge), scientists can carry out their experiments in the model. A similar model is being developed for the San Francisco Bay area. The models will simulate various attempts to reduce smog, such as showing exactly where and how reductions in automobile traffic would improve the air. "A model might show us that an automobile cutback would help in San Francisco, but do nothing for [nearby] San Jose," Gloria said. "Obviously, models are a more convenient juncture for finding defects in a plan than in actual practice. The first California smog devices on cars in 1964 might never had been introduced if modeling had been available. Those devices on the cars have *altered* the smog, rather than reduced it. Hydrocarbons were reduced, but ozone actually increased as a result of them."

By using the NASA immediate-sensing aircraft to develop models, and by simulating various smog plans in the models, truly effective plans may be developed in the next few years. Questions can be answered before new concepts are tried and, sometimes, found lacking. What would be the impact of a new industrial park in X location? What would be the total effect of replacing a million cars with five hundred short-haul aircraft? These kinds of proposals, being made today with increasing frequency, deserve careful model analysis before being cranked into an expensive pilot program in real life.

Remote Sensing

The smog monitoring necessary for developing models, which is today being handled by the immediate-sensing aircraft, might eventually be undertaken by satellites, with models being developed on a much broader geographical basis. In order for satellites to identify the air constituents, however, more precise remote sensors would be needed than those being carried by ERTS. Sensors would have to be equipped to divide the spectrum into very narrow bands in order to distinguish the signatures of various air constituents. The four ERTS channels are too broad to allow such identification of the composition of air.

"However, I see no technical reason why we cannot equip satellite sensors with narrow band widths and identify a number of air constituents," says Dr. Joseph Behar, of the air center. "Remember, astronomers have been identifying gaseous matter around distant planets for decades."

"They've been identifying it, but who knows whether they have been right?" we asked Behar one day at Riverside. After all, it's not easy to obtain ground truth on Saturn.

"True," Behar conceded with a begrudging smile. "We have no indisputable way of identifying matter on other planets. But we do have methods on earth. Airplanes could offer the ground truth for satellites."

ERTS Air Project

Such experimentation, combining satellites with aircraft, is probably still some years off in smog work, although some basic study has begun with ERTS. Every eighteen days, at about 9 A.M., Dr. Ernest Rogers of the Los Angeles-based Aerospace Corporation drives with his wife Eda to the Torrance airport a few miles from their home. Eda is there to rent a single-engine aircraft and log a few hours of flight time. Rogers is there to ride in the right seat and make observations of the haze over the Los Angeles basin, as part of an ERTS project.

While Eda is going through her outside-cockpit checkoff procedure, Rogers is standing nearby, peering toward the sun through a telescopelike instrument known as an aureole, which records the haze-intensity level of the sky. Once in the air, the Rogerses climb above the haze layer, subjectively observing the haze pattern and roughly charting it for eventual comparison with the Los Angeles picture being obtained by ERTS at that very hour. Their subjective in-flight comparison, later combined with simultaneous aureole readings being taken by the Rogerses' colleagues on the ground, together will be used as ground truth by Rogers in evaluating the success of ERTS in detecting that morning's haze.

While haze has been observed from space since the earliest Gemini photography, an evaluation is needed to determine if satellites can detect it when only a thin layer is present. The degree to which ERTS can detect these thin layers, distinguishing the edges of the coverage, must be established. The appropriateness of the ERTS 9:30 A.M. sun angle is also to be evaluated, as are the relative merits of the four ERTS bands. The haze tends to blend in with the ground in the relatively broad frequency bands used by ERTS, again a reason for using narrower bands in some future experiment.

"We're taking a basic first step in studying smog patterns from satellites. Eventually satellites may provide a daily coverage of a broad area, and we would hope to track fog and smog as it progresses from one area to another."

At present, no satellite sensor has been used to distinguish between smog and clean fog. "That has to be the next step," Rogers acknowledges. "Right now, we are simply detecting the aerosols in the air, whether they be moisture or pollutants."

The ability to map the haze layer, whether fog or smog, would have some immediate value as weather information for aviators, since geosynchronous satellites have not yet been able to detect thin layers of haze for daily weather reports. However, the primary objective of haze study is strictly to obtain information for air pollution control. Rogers' basic study can lead the way to sophisticated work with greater repetition, broader area coverage, and more critical band width.

The Need to Look Sideways at Smog

While future experiments from space may improve upon the ERTS results, the satellite overview will always be at a disadvantage when compared to the viewpoint of the immediate-sensing airplane. The aircraft flying within the haze (as in the NASA-Riverside experiment) can stratify it into layers, a fundamental requirement for good haze modeling. An over-all movement of a bank of smog might be plotted from a satellite, but help would be needed from an aircraft to determine the altitudes and thicknesses of particular layers—unless, of course, satellites find some way to look sideways at a cross section of the earth's weather.

Actually, some random experiments doing precisely that have already been accomplished by both NOAA and Department of Defense satellites, and a comprehensive project aimed at stratifying layers of air around the earth has been proposed by a NOAA atmospheric physicist—named, appropriately enough, Walter Planet.

Planet's project might occur this way: In 1980, an astronaut in a Space Shuttle spacecraft would prepare for an experiment by aiming a narrowly banded sensor just above the earth's "limb," or horizon. It would be nighttime for the astronaut, in that the earth would be positioned between the sun and himself, providing him with a solar eclipse. As the eclipse would end, and "sunrise" would occur for him, he would train his sensor on the sun as it first peeked over the earth. He would vary the wave lengths and filters on his equipment, tuning it finely in narrow bands of the spectrum, and elevating his lens to examine different altitudes above the earth's horizon. The result would be a stratification of the earth's haze layers which might add considerably to haze modeling.

Strange Problems in the Stratosphere

But besides using stratification to study the earth's troposphere, Planet sees another use for it which he regards as even more significant—and which can perhaps best be appreciated by examining the roots of his concept, beginning in New England in 1816.

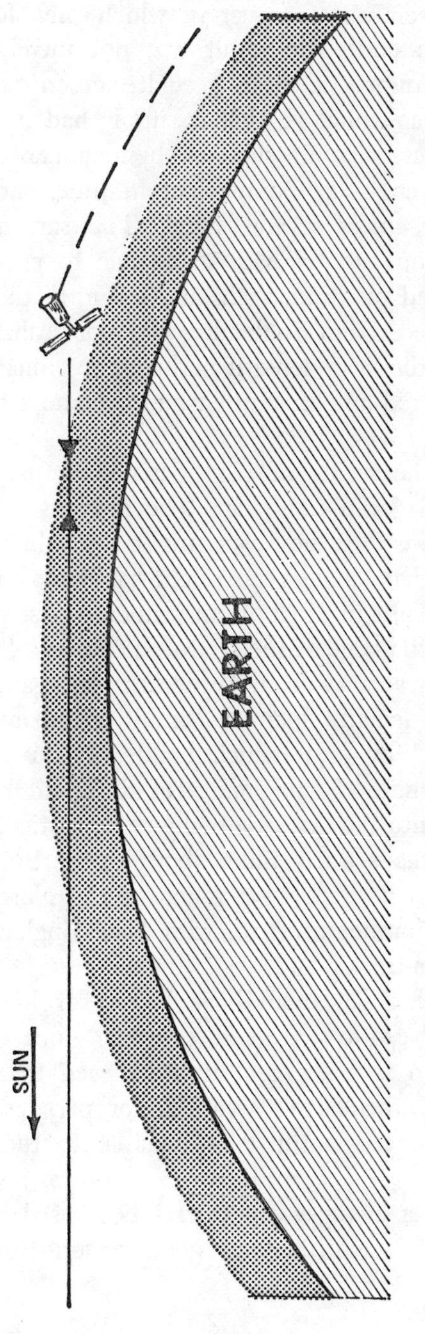

Figure 8: Certain future satellites may sense not the earth, but its atmosphere, and provide either endorsement or condemnation of the SST and other concepts.

Eighteen sixteen was the year in which the New England winter forgot to end. The result was not merely an unseasonably cool summer; a more accurate description of New England that year would be that it simply had no summer at all. Temperatures were in the 40s by day and the 30s by night. Blizzards engulfed Connecticut in June, and six inches of snow spread across several states. The sun never shone brightly, and crops simply had no chance to grow, although farmers continued planting, again and again, as the "summer" months continued. New Englanders did not realize what was happening, but they wondered if perhaps the climate had been dramatically altered and if summer had become a thing of the past.

Fortunately, the spring of 1817 dawned normally, and the nightmare of 1816 did not recur. The factor which scientists say caused the weather, however, might easily have continued for more than that single year; scientists blamed the absence of warmth and light in 1816 on an immense dust cloud which accumulated in the edge of the stratosphere over New England, as a result of several volcanic eruptions thousands of miles away. They particularly blamed Mount Tambora, an island volcano just east of Java, which had exploded in 1815. An estimated one hundred cubic miles of dust had been spewed from Tambora into the sky, eventually to be carried up into the stratosphere, perhaps thirty miles above the earth. From there it apparently cast a shield over much of the Northern Hemisphere in that summer of 1816, with New England somehow receiving an extra-thick cloud.

Results of other volcanoes have lasted for longer periods. Krakatau caused exotic sunsets all over the globe for a decade after its eruption in 1883, and it is believed to have caused some cooling of the earth's weather, for perhaps one year—although it produced nothing comparable to the 1816 summer.

If such weather changes can in fact be caused by a canopy of volcanic dust, scientists have good reason to become interested in the stratosphere. But since man has no control over volcanoes, the interest would seem to be more academic than useful.

Man Affecting the Stratosphere?

However, scientists have, for the last two thirds of a century, been accumulating evidence that volcanoes may not be the only source of these outer-space accumulations of dust. Another contributor to the phenomenon may be man himself. Man is being implicated by some evidence pertaining to the earth's albedo, that fragile heat budget that keeps the earth's temperature in balance. Since all of the earth's energy comes to us from an outside source, it must pass in through our stratosphere and troposphere, and all the heat we discard must pass outward through these same layers. If the amount flowing out is different from that flowing in, thereby throwing the albedo slightly out of balance, strange things can happen in both the short and the long range. From 1910 till 1950, for example, the earth experienced a warming trend. Physicists eventually developed the greenhouse theory, and attributed the problem to man's industrial pursuits. The CO_2 given off by factories, so the theory goes, has created a canopy in the earth's stratosphere which allows solar IR radiation to pass in toward the earth, but which seals in the longer-wave far-IR emitted from the earth. This would increase the entire world's temperature, with some pleasant short-run effects in regard to winter environment and crop seasons, but would also bring in a number of new headaches—the melting of polar ice being one troublesome prospect.

Since 1950, however, the world has begun to experience the opposite effect, a cooling trend. The particulate matter in the air has been generally suspected as being the culprit, behaving much as the volcanic dust in 1816, and while obviously not yet that severe, it could be gradually introducing unpleasant consequences on a global basis. In most cases we would find the cooler weather less agreeable, with crops becoming increasingly difficult to grow in the higher latitudes. First the grain belts in Canada and then our central states would have their seasons cut too short for crops to mature. Finally a heavy accumulation of ice at the poles would occur, and another ice age would eventually overtake us, although that extreme would not be reached until eons hence.

The SST—Health Detriment or No?

Physicists had noticed these trends for decades and talked with one another about them, as scientists do, with no one else becoming terribly excited. But then, in the 1960s, the age of ecological concern germinated, and flourished. It was reaching full bloom when the Super Sonic Transport plane reached its moment of truth, and attention from the Congress was focused on some of the alleged dangers of that project. The fact that each SST flight would unleash tens of tons of dust into the stratosphere, and hundreds of tons of water vapor, was a truth accepted by all. But what unfavorable (or favorable) effect these quantities would have on the stratosphere was a matter still in question. One school of scientific thought had it that the effluents released by the SST would significantly reduce the layer of ozone in the stratosphere. This, they maintained, would allow a great increase in the UV radiation passing into the lower atmosphere. The medical world had already become convinced that skin cancer was most prevalent on such human bodies as those displayed by perennial sun worshippers around tropical resorts. Now, with the SST, some of the same doctors feared that the UV radiation would become great enough to create skin cancer on almost all earthlings, simply as a result of everyday encounters with sunlight. They visualized an environment where a short stroll from your parked car into the office or supermarket would allow your skin to be burnt, making it dangerous to be outdoors without an umbrella.

During the SST hearings in the Congress of 1971, scientists

were paraded onto Capitol Hill by lawmakers on both sides of the SST controversy, testifying either that the SST pollution would or would not cause harmful effects in human beings. During the course of the proceedings, it became painfully apparent to the public that (1) the question was indeed a matter of critical importance, and (2) we didn't know the answer to it.

The pro and con arguments quoted in *Congressional Digest* are interesting. (Pro) "According to existing data and available evidence, there is no evidence of likelihood that SST operations will cause significant adverse effects on our atmosphere or our environment . . . *the desired degree of certainty about these matters has not been attained* . . ." (Con) ". . . the possible damage to the upper atmosphere has received inadequate public attention . . . possible effects on weather and climate *are not well understood* at this time . . . With respect to the destruction of ozone and the consequent increase in ultraviolet radiation, *little is known at this time*. . . .[1] (Italics added.) The two arguments, when addressing themselves to scientific evidence, are strikingly similar in their frankness about the lack of information. In the absence of fact, some congressmen resorted to emotion. Badgering of scientific witnesses to make their opinions seem invalid went to the extreme, and an uneasy atmosphere surrounded the environmental issue.

There were, of course, considerations other than environmental relative to the SST, for both sides. Economics, employment, balance of payments, national transportation priorities, and the subsidizing of industry all figured prominently in testimony. But these were not new arguments. They had existed for the ten years that Congress had supported the SST and had been much in evidence during the previous session, in 1970, when the House three times had voted in favor of the plane. The difference in 1971, as stated by several of the congressmen who changed their vote between the two years, had been the uncertain yet frightening cloud of environmental catastrophe which had built up in 1971 scientific testimony. The final votes against the SST were 51–46 in the Senate and 215–204 in the

House. The closeness of both these votes was just one of the factors which seemed to underscore a lack of clear-cut environmental information regarding the SST.

The Congress of the United States can be commended for examining the environmental issue, whereas France and the Soviet Union moved unflinchingly ahead with their own SST plans. The unfortunate aspect of the situation is that science provided Capitol Hill with so few facts. Congress, with a good deal less evidence than in the matter of Los Angeles smog, had rightly or wrongly succeeded in tightening the reins on the acts of man, or at least of Americans. But the matter has since appeared to be far from resolved, with Washington sources periodically predicting that the United States will eventually have its own SST.

The SST, as we have already seen, is only one of the various impacts which man might be imposing on the stratosphere and which could affect us, perhaps fairly soon. The need to have a better understanding of factors affecting the earth's albedo is as essential as learning about the bacteria affecting the temperature and health of the human body. Unfortunately, we are more limited in our understanding of the world than we were in our knowledge of the human body a full century ago.

This void in scientific knowledge was the real reason for Walter Planet's proposal to NESS calling for a study of air quality, *especially including the stratosphere.* Planet's Space Shuttle astronaut, whom we left a few pages ago looking at the troposphere, might now be instructed to aim his sensor slightly higher above the earth's horizon and study our planet's stratosphere. In his sensor he would use a series of narrow bands covering most of the spectrum and identifying the air's constituents. Among other things, this sensing of air quality would establish an ozone molecule count and a measurement of water vapor. These and other factors could be checked repeatedly over a period of years to determine if and how the stratosphere would be affected by man's activities.

The astronaut and his ability to judge results and direct his search accordingly would have certain advantages over an unmanned satellite at the outset, although if the experiment proved meaningful, Planet would expect eventually to utilize a permanent unmanned satellite, sensing in selected bands.

Ideally, the first such observation would have been made before any SST flight had occurred, but inasmuch as French and Soviet SST programs are moving ahead rapidly, and since by the 1980s we ourselves will have been making occasional supersonic bursts with smaller military SSTs for twenty years, this is obviously impossible. (The physicist, for that matter, would like to have monitored the air in the stratosphere before the Industrial Revolution began, but that, too, would now seem difficult to arrange.) However, by observing the spectrum repeatedly as the SST flights continue, physicists could determine if an effluent from the jet discharge is indeed gradually reducing the ozone.

The results might at long last clear the way for a then less controversial American SST, or on the other hand might convince the French and Soviets that they were pursuing a catastrophic course for all mankind. Planet seemingly has no preconceptions about the SST, having commented that the 1971 conjecture on both sides was premature. And he points out that while the SST prompted his proposal to NESS, jet emissions are only one of the factors possibly affecting the stratosphere and our future. "Could it be that particulate matter in the stratosphere is halting enough of the solar rays to cause the earth's cooling trend? If we keep this up, are we in for increasingly long winters?" asks Planet. "We think it might be wise to answer not only these questions but some which have not yet even occurred to us as well." Both the asked and unasked questions may have critical answers awaiting man's discovery in the stratosphere. The earth's albedo should presumably be under constant observation for early signs of crisis.

The satellite programs, incidentally, include few, if any, scientists who see man as a dinosaur in his final chapter on earth. What they do see is a planet that needs an outside view with systematic observations, leading to early guidelines for survival. The New England episode in 1816 should suggest to the most skeptical that the outside supply line of Spaceship Earth can perhaps be blocked. The temperature variations and SST emissions are enough to convince us that the acts of man, and their effects, require repetitive observation, be we healthy at this point or not; earthlings must assume the responsibility of monitoring their present and future activities on Spaceship Earth.

Man's Role with Hurricanes

We look for satellites to inform us whether or not the stratosphere is truly collecting a canopy of man-made waste. Already, however, we recognize the fact that we are cluttering up the closer, tropospheric band of air that holds the earth's daily weather. In view of man's constant, if inadvertent, changing of the environment with his smog, it seems inconsistent that he is hesitant to take *deliberate* steps to alter the weather.

Hurricanes are a striking case in point. In 1972, floods and tornadoes in the East were the latent products of Hurricane Agnes. In June, Agnes had ripped north through Florida toward Washington, D.C., and the Northeast, where she then spewed forth all the water she had sucked up from the Gulf. During the flash floods which Agnes created, flood fighters and evacuators had to race the clock in desperate attempts to save people and property. In some cases the clock was the victor; hundreds of lives and more than a billion dollars in property were lost because more than once during the crisis man had run out of time. And yet Agnes, the week before, had offered man days upon days to pursue her at sea; if she could have been dissipated then, before she had taken that enormous drink from the Gulf, one of the worst floods in the U.S. weather history would have been avoided.

Agnes indeed had enjoyed a long and leisurely life in the world of weather phenomena. In early June, she had been conceived off the coast of Africa, the child of a warm breeze and a low-pressure trough. She had gestated for several days across the Atlantic, one of many ripples under the wary eye of ATS-3. She had then been born into the world as a tropical storm near Yucatán, Mexico, June 14, and had started life with a northerly run toward the United States. The warm waters of the Gulf had stimulated a fury that was in her, and her development had quickly earned her the designation of Agnes, first hurricane of 1972.

At this point, two days before she was to strike the Gulf coast, we earthlings knew full well what to expect. After all, man for generations has known a number of vital statistics about the perennial "hurricane witch." He knows she typically stands 40,000 feet above the sea and is surrounded by a hundred miles of swirling bands of pouring rain. He knows that her lesser gale-force storms spread over perhaps 400 miles. He has learned that at her feet, in the first mile above the sea surface, ferocious air and hundreds of tons of water charge in toward her center to be pulled upward and stored as ammunition. And he has calculated that her wrath produces as much energy, in a single day, as the fusion of four hundred twenty-megaton H-bombs. Furthermore, he knows full well that in a few days this enormous force could be unleashed on him and his home.

Yet this same man who has spent his entire history throwing spears or cannonballs or bombs can only spend several days *observing* the storm—instead of reacting as he did at Pearl Harbor or in the war against poliomyelitis. A few stouthearted hurricane hunters venture out to the witch in airplanes to check on her condition, taking her temperature and picture, among other things, as part of the waiting game.

But we wonder if this business of taking the pulse should not give way to a practice of actually fighting the storm. The art of cloud seeding is not yet fully proved as a means of neutralizing a hurricane, but two attempts have turned out to be fairly successful. The seeding technique is to fly over the storm, just outside the warm eye and drop explosive canisters of silver iodide into the cold swirling updrafts. This seeding, repeated frequently, is designed to crystallize the super-cool water droplets of the storm and warm this ring of cold air. As soon as the eye is surrounded by a warmth almost as great as its own, it expands to include the warming winds closest to it. As the calm eye is enlarged, a ring of several miles of violent updrafts is quelled, and dissipation of the hurricane has begun.

There is general belief that this or some other form of seeding can be effective, although the technique is still in question. Scientists, characteristically enough, say the primary need is for more data. So each year, in Operation Storm Fury (a federal multiagency effort), weathermen and military flight crews experiment with those few Atlantic hurricanes which they are confident will *not* strike land. NOAA's theory is that experimental seeding should be conducted only on storms that will remain at sea, just in case the whole experiment backfires and the witch becomes angrier and potentially more devastating than ever. Gulf hurricanes, which inevitably strike land, are therefore not eligible for Storm Fury. The seeders wait each year for appropriate storms in the Atlantic, of which there are few. For some unexplained reason, the fertile Pacific laboratory where the *chubascos* are numerous, and usually located safely at sea, is also ignored.

So the experimental seeding of hurricanes, which was begun in seriousness more than ten years ago, moves ahead at a leisurely pace. The improved detection ability and data base offered by GOES pictures will conceivably speed up this project in the mid-1970s and help to control the hurricanes which now tromp over North America at will.

Dr. Joanne Simpson of the Hurricane Warning Center in Miami, appearing on the "Today" show in 1972, pointed out that "we don't want the hurricane, but we want its moisture."

Whether or not flood-drenched residents of Harrisburg, Pennsylvania, following Agnes' assault, would have agreed with Dr. Simpson is doubtful. *Some* of the moisture would be welcomed, and weathermen say the trick is to dissipate the storms partially at sea, allowing them to come ahead and bring rainfall ashore, but in a less concentrated dose.

Changing the Daily Weather

"Weather modification" is the label the government has hung on the cloud-seeding activity, and it pertains to numerous weather phenomena other than hurricanes. Earthlings have considerable potential (and many limitations) in reorienting the more destructive forces of Nature. While man cannot *create* weather, or make rain out of dry air, he can modify it by augmenting natural processes. Much as a doctor can help an expectant mother avoid miscarriage at one time, and yet can induce labor at another, man can dissipate a storm on the one hand and trigger it on the other—provided Nature does indeed have a baby available to be born.

On the West Coast of the United States, satellites have contributed considerably in the detection of approaching Pacific coastal fog, and science has also learned that these cool fogs can be cleared by a seeding, temporarily producing snowflakes never to strike the ground. It is so effective a process that a few minutes after an airplane has seeded a specific sector, that area becomes a distinct hole in the blanket of fog. This expertise will be used on a broad basis if the U. S. Department of Transportation has its way, that department being perennially unhappy about the way fog ties up airports, harbors, and freeways.

The Forest Service, with its interest in reducing forest fires, has as great an aversion to lightning as Transportation has to fog. Smokey the Bear has for years been saying, "Only *you* can prevent forest fires," but the facts would seem not to support him. As an example, on one August day in 1972, more than a hundred fires started from lightning in Idaho. The ability of satellites to locate cumulo-nimbus clouds early in the development of electrical storms has already proved valuable. Satellite pictures can be used to locate the clouds and allow aircraft then to move in and seed them before they can produce storms and lightning.

Considerable enthusiasm also exists for the elimination of hail-producing storms in such areas as northeastern Colorado. Sugar beets and corn crops in that area are often completely destroyed by heavy hail, and insurance is correspondingly prohibitive. A program of seeding hail clouds has apparently done a good job in Colorado, avoiding about ten dollars in probable crop and property damage for every dollar the government has spent, according to a NESS spokesman.

The cloud action following the seeding, of course, has been recorded on film by the regular ATS—now GOES—pictures. The satellite pictures offer an overview of exactly where the clouds go and what they do, revealing that precipitation may take place elsewhere, for good or for bad, as a result of stopping hail in Colorado, and the satellites can silently record the whole sequence. This may prove important, since precipitation has proved considerably easier to induce than to reduce.

Rainmaking is a good example of inducing precipitation. The seeding of rain clouds over drought-stricken areas is practiced on an experimental basis in much the same manner as in a hurricane. Again, satellite imagery records not only the action in the rainmaking area but the results that may occur for long distances downwind. The seeding of clouds in Missouri will perhaps affect the weather in Ohio, and only a satellite offers the perspective to observe both.

Snow blizzards have been successfully—and controversially—averted in upper New York State by seeding clouds and reducing the size of the snowflakes. The smaller snowflakes are carried with the wind, falling over a wider area instead of unloading an entire blizzard close to the Great Lakes, where Buffalo, New York, has long been a favorite dumping ground for snowstorms. Ski resort owners have complained that cloud seeding has shortchanged the snow on their slopes, while weathermen have insisted that the reverse has been true. How can their argument be solved? Constant observation of the area with GOES's elapsed-time pictures could keep both the weather modifiers and the public aware of what truly is being done, and to whom.

Weather-modification practices seem certain eventually to

become a part of our lives. In some cases they will save tens of thousands of persons from diabolical disaster, at little if any sacrifice to man or Nature. In other cases, numerous arguments, either commercial or environmental, may arise against the process. In such cases, the satellite, by providing a documented over-all view of the matter, will offer man an opportunity to take action based on intelligent judgment.

These "acts of man" which affect the weather, along with man's other acts regarding air quality, should—each time we think of the weather or draw a breath of air—emphasize for us the role of satellites. The control of our weather, and of the air immediately around us, and even of the canopy of dust in the stratosphere, lies first in understanding such phenomena as those which we have superficially discussed in this chapter. Satellites would seem to offer the observational tools critical to gaining much of this vital understanding.

13. Man and Water

Each summer for the next several years a scene will take place at a number of inland lakes in the United States.

A vacationer will have decided that he and his family of four can afford to return to the delightful lake that they discovered several summers before. Arriving in mid-afternoon, they will jump into their swimsuits, alive with anticipation. . . .

"Last one in is a rotten egg," shouts Offspring No. 1, and they all charge out to dive into the lake.

The sprint to the beach is fast at the beginning, over the familiar wooded area. But then, in the last hundred feet, the four pairs of scrambling legs suddenly shift into slow motion as their owners see what lies ahead in the water.

"You'll be a rotten egg, all right, if you jump in there," shrieks Offspring No. 2. Rotten eggs, sure enough, would seem already to be in the lake. The first several feet of water are choked with weeds and scum. The water is green, not blue.

Our hypothetical family looks up and down the beach. It's more of the same in both directions—green swamps instead of blue water.

By now, their bewildered reaction is likely to be followed by harsh words, chosen to describe the deteriorated waterline. A word the family might overlook, however, is the term biologists have foisted upon us in recent years: "eutrophication," which refers to the choking of lakes by plant life. Lake eutrophication is one of several obvious water-pollution problems confronting the United States and much of the world. While perhaps a third of the United States faces shortages of water, as described in

Chapter 9, all of us face a shortage of *clean* water. Lakes, streams, and coastal waters that used to be relatively blue are accumulating the green of algae, the brown of sediment, or, occasionally, the black of petroleum.

Remote sensing can bear a role in identifying the often illusive pollution sources, and in fact both aircraft and ERTS have already begun to make their mark, as we will see in this chapter.

The Choking Lakes

Eutrophication is a phenomenon in the life cycle of lakes which would occur without man ever contributing to it. Nature, however, if unassisted, might take thousands of years to choke a lake to death with vegetation. Man, on the other hand, with his life style of the mid-twentieth century, is capable of speeding up the process to ten or twenty years.

The Federal Council on Environment Quality, in its 1972 report, indicated that lake eutrophication was a phenomenon streaking to super-polluter status, and blamed land runoff from farms (fertilizers, pesticides, etc.) as a major cause. The algae which we so eagerly sought in our search for fish at sea is often *persona non grata* in lakes, where the water cannot support both the algae and the chain of fish life that consumes it. That being the case, much of the algae drifts along uneaten, sometimes becoming entangled in the standing weeds of the shallow water, which similarly are flourishing as a result of the inflow of fertilizers and such. The water is overenriched.

Even such states as Minnesota, the "land of 10,000 lakes," and Wisconsin, "the land of sky-blue waters," have lost their chastity in recent years. With tens of thousands of lakes in the Midwest alone, there is a need to identify the lakes that are in trouble and the factors that have brought it about. Where is the pollution originating? Which lakes are decaying? Which lakes are sustaining a reasonable growth of algae, capable of maintaining good fishing stock? Which lakes have poisonous algae?

The University of Wisconsin has been flying sensing aircraft over nearby waterways for several years. Biologist Dr. James Clapp of the university says that biologists have been able to obtain color images which, when spectrally analyzed later in the lab, could be separated into several distinct varieties of algae. The sources of pollution become readily apparent in such laboratory analysis; furthermore, the degree of eutrophication can be effectively measured, indicating comparative water quality on a lake-by-lake basis.

This is the kind of information that can be provided effectively by an airborne sensor, and by no other way, says NASA biologist Dr. Ellen Weaver. "In Clear Lake, California, chlorophyll readings taken by boat are more than ten times as high on one side of the lake as they are on the other. The boat gets a reading of five milligrams per meter on the south side in the morning. By the afternoon, when the reading is taken on the opposite side, the wind has caused an upwelling and the readings may have shot up to more than fifty. The way to accurately map the pollution patterns is from an airplane, quickly reading the whole lake at the same hour."

Once a lake's pollution pattern can be modeled, the solution may lie in treating the problem itself with phosphate-reducing chemicals. More likely, the answer will lie in going to the source and correcting what may be only a careless example of soil erosion on a nearby farm. If a fertilizer-laden field can be identified as the source, the farmer could be instructed to commence erosion-preventing tillage practices. Progress of the recovery of each lake can be monitored regularly and accurately; the high reflectance of certain algae can very directly indicate the phosphorus content.

The University of Wisconsin work has included the monitoring of sediment in lakes and streams near paper mills. The researchers have located plumes of paper-mill waste; have found ways for aircraft sensing to make in-water pollution detectors more effective; have provided data about paper-mill emissions to mill owners who were conscientious, and have used it to take legal action against those who were not—Wisconsin and Florida being the first two states we know of in which remotely sensed (aircraft) data was successfully employed in water-pollution cases.

The Rivers

Looking at San Francisco Bay in Chapter 9, we made reference to sediment as a surrogate signature for fresh water. San Francisco Bay is only one of a dozen estuaries, rivers, and bays that are receiving attention from sedimentologists, those experts on mud who are studying ERTS data. Problems vary considerably from locale to locale. We saw how the absence of freshwater sediment was viewed with alarm in the California delta, where the idea seems to prevail that a muddy stream is a healthy stream.

Mud in Mississippi River water, however, generates considerably less enthusiasm. In this longest of all rivers, water is consumed and reconsumed as it passes downstream by the river-town residents of seven states. A glass of water in New Orleans may have already been drunk by a series of two dozen persons along the Mississippi's meandering path; microorganisms multiply prolifically in this situation. Mississippi mud (sewage or runoff from farms) is, like the lakes of the nation, full of nitrates, phosphates, and other fertilizers. Once the fertilizer has entered the water, and been consumed by algae, a diet is created for zooplankton, unicell creatures which multiply rapidly on the algae, all the while stealing oxygen on which fish might otherwise thrive. Thus turbidity becomes a benchmark of a chain of life in the water—a chain which often expands so quickly in its second or third link that it cannot be supported; fish, again as in the lakes, simply disappear.

One body of water where the sediment is not only under the sensors of ERTS, but directly under the eyes of Washington, is the "Nation's River," as Lyndon Johnson good-humoredly designated the Potomac a few years back.

A hundred and fifty years ago, when colonists were scooping fish out of the estuary with their hats, Johnson's designation would have been apt indeed. But in recent summers, when the aroma of decaying algae has drifted downwind to the Capitol Mall, our nation's leaders have been less than proud of their official child.

The odor of the river, however, has not yet stimulated meaningful action. In spite of the proximity to lawmakers and environmental administrators, no solution has yet been visualized to return the Potomac to its more fragrant days of yesteryear. Instead, sediment continues to be trapped in a river whose bottom since Revolutionary times has been piled ever higher with chronicled layers of back-yard history.

Federal figures released in 1972 indicated that in the previous year more than five thousand additional streams had succeeded in qualifying for the government's designation as "polluted." The challenge in solving these problems in often remote streams is staggering, particularly when we realize that the much more accessible "Nation's River" has not been improved. Comprehensive action has been slow to come because, while the Environmental Protection Agency (EPA) has been painfully aware of the general causes, the agency has found the precise sources of pollution evasive.

ERTS Locates Pollution Sources

Treated and untreated sewage alike have contributed phosphates and nitrates to the Potomac, and the locations of sewage outlets into the river are obviously known, American University's Dr. Norman Macleod commented to me in June 1972. This is not true of pollution sources other than sewage, he hastened to add. Sediment rich in fertilizer was being washed into the river from farms where the fields have been allowed to erode; mud was being washed in from bare construction sites; tailings from mines were streaming into the Potomac, with their share of sludge and minerals (which, conveniently, respond to remote sensing with unique metallic signatures).

Later that year, ERTS, with its repetitive overview of the entire basin, helped Macleod and others put the environmental jigsaw puzzle together, providing sedimentologists with pictures of the Potomac at eighteen-day intervals. They could now begin to monitor the eternal progress of pollutants in all seasons and in all climatic conditions. "The September 23, 1972, ERTS picture showed the low-water sediment," said Macleod. "Mostly organic pollution. Then the October 11 picture, which followed a heavy storm, showed plumes of heavy sediment flowing in from West Virginia and Maryland."

The satellite picture shows the extent of the problem—starting with the bare spot of ground where the erosion begins and following along to the place where it flows into the river as a brown plume. "ERTS has pointed out sources of pollution precisely in several instances," Macleod says. "A Maryland sand-and-gravel plant in one case. Some open gravel areas in another. And some shopping-center developments."

Some of the purveyors of pollution are unintentional, others are flagrant offenders, but in either case they are usually breaking the law. (Farmers are required to practice minimum tillage; contractors are limited in the length of time in which they can leave bare ground unprotected; mines have disposal requirements to meet.) These problems, often remote and hidden in the rolling hills of the countryside, frequently go unnoticed on the ground. "It has been difficult for authorities to patrol their areas. Prince Georges County [Maryland] has only three persons who perform inspections," said Macleod. "And one of the offenders we have noticed in Maryland is not at all visible from the road," he went on. Macleod's pollution studies, however, were made available to local authorities, giving them the opportunity to "see beyond the road" and enforce the regulations.

In some cases, practices which are causing water pollution *are* within the law. But the remote-sensing overview will now undoubtedly become useful to legislators everywhere looking for ways to regulate water pollution at its usual source—on the land.

ERTS Discovers Sewage Violators

Macleod had told me in 1972 that sewage outlets into the Potomac were already known. But after data from three ERTS passes had been collected, it became obvious that this was not entirely true. Data from the ERTS passes in September and October 1972 and January 1973 was analyzed for selected test areas in the Potomac and other nearby rivers. The analysis work was accomplished, not merely with visual images, but with digital tapes which quantified the river pollution in a computer model at NASA's Goddard Space Flight Center. The object was to pinpoint sources of pollution.

Rather astonishingly, the information from 570 miles in the air located sewage effluents which had previously gone unnoticed in the water. Subsequent water sampling in the rivers themselves proved the ERTS tapes to be providing accurate information, and the data was translated into action; after ERTS traced an unregistered (and untreated) sewage dump to an apartment complex on the Patuxent River, the Maryland Environmental Health Department issued a compliance order, says NASA's Dr. Jane Schubert. Similar data, says Dr. Schubert, is currently being used as the evidence in a 1974 suit by a citizens' committee filed against the Washington Suburban Sanitary Commission. The commission is charged with releasing improperly treated sewage. In all these various actions, ERTS is providing the initial detection information, with details as to bacteriological content of the effluent necessarily collected subsequently on the ground. The ability, demonstrated in our Wisconsin example, of sensors in aircraft to analyze pollution by variety is a refinement not yet attained by satellites. But the simple ability to detect violations on a broad basis was impressive enough for EPA administrators to show ERTS color pictures to Congress in justifying their water-quality budget requests for 1974.

Synoptic Look at Eastern Waterways

The superiority of aircraft sensing in certain respects, such as in the ability to identify algae, is readily apparent. Could not all this gathering of information about the Potomac River basin have been better done by aircraft? Certainly the resolution from an aircraft would have provided a number of details not possible with ERTS. And unlike some vast areas we have discussed in previous chapters, the Potomac basin is small enough so that one might expect an airplane to cover it reasonably well.

The expense of aircraft surveillance, however, especially on a repetitive basis, would have been considerable. The cost of ERTS, on the other hand, is justified by satisfying the requirements of some three hundred projects, including a dozen hydrological investigations. Quite beyond these formal investigations, however, the wide sweep of ERTS uncovers many an unexpected piece of information. The New York garbage ship, quietly offloading its cargo just a few miles off the Jersey shore, is one dramatic example. Its acid wake was picked up by an ERTS picture in August 1972. Water authorities in the greater New York area, whether or not they are sponsors of ERTS projects, will be well advised to keep abreast of satellite data, for more reasons than one. The Hudson River has salt-water problems to rival those we saw earlier in San Francisco Bay and it "may well be the most complicated estuary in the United States," according to at least one Army Corps of Engineers spokesman.[1]

Furthermore, such eastern lawmakers as New York Congressman Hugh Carey, responsive to water-quality problems, have expressed infatuation[2] with California's seventy-five years of hydrological adventure (already discussed in Chapter 9) and decided that the answer to water quality lies in such Roman-style diversion systems. There have been a number of proposals since 1966 involving eastern rivers. The Army Corps of Engineers has figured in many of the proposals, but independent concepts in various states have been made without consideration one for another. Creative thinkers have made plans for most of the waterways along the Eastern Seaboard which have varied from damming up Long Island Sound to skimming Lake Champlain. Proponents of the systems have pointed to better distribution of water, utilization of marshlands, substitution of fresh water for salt water, and control of floodwaters. Critics, on the other hand, see the multiplicity of tunnels, dams, and impoundments as a sort of hydrological overkill, creating traps of algae, mishandling of sewage, and destroyers of wildlife and scenic beauty.

So far as we know, neither proponents nor critics of the various proposals have sought satellite information. Yet proposals relating to diversions of water involving the Hudson, Housatonic, Raritan, Passaic, Connecticut, Delaware, and Susquehanna rivers, Lakes Champlain and Ontario, and Long Island Sound can obviously not be understood from purely local studies. An ERTS overview of the entire area now being available for the small price of the pictures, we wonder if this might not offer a giant first step toward resolving the problem.

The "Good Spring Water"

Vermonters, incidentally, have not been a party to the various legislative schemes to skim their lake into the water pipes of neighboring states. Although the Lake Champlain of the 1970s is somewhat less than crystal clear, it nevertheless ranks as one of the brighter jewels in the Vermont diadem, and the residents have their own ideas about its use. In fact, a comprehensive ERTS study of the lake and its basin has been undertaken at the University of Vermont by Dr. A. O. Lind, a geographer, along with faculty geologists and engineers. Lake Champlain having some of the pollution problems discussed elsewhere in this chapter, Lind's group is making a study of its entire circulation system.

The unique part of the study, however, is directed not toward the lake itself, but toward the nearby Green Mountains. The mountains are the source of most of the water draining into the lake, some of it above ground, some below. The project objective is to find the underground water, the "good spring water" Vermonters like to talk about. Unfortunately, some of the spring water is suspected of being contaminated by septic tanks along its path to the lake. But since the underground springs tend to meander unpredictably, no one knows which septic tanks might drain into the water.

Investigators began their search for the underground springs with the first ERTS images in September 1972. Lind and his co-investigators noted the sprinkling of vegetation on the imagery, the summery green foliage of the maple, beech, and birch showing up bright red on the pictures. In subsequent ERTS pictures, as the leaves lost their chlorophyll and turned to brilliant autumn yellows and reds, the colors on the imagery did just the reverse; by the time of the early October imagery, the vegetation had faded to a dark pink.

In some places the vegetation stood out in masses of color, but in others the pinkish lines ran through the mountains like rivulets. Some of the lines *were* streams (or, rather, the trees along the stream banks), winding their way, above ground, down into the Champlain valley. Others, the investigators were hoping, would turn out to be fault lines.

The presence of faults, and their significance, became more apparent as additional images were studied throughout the autumn. By the time the leaves had fallen in late October, Lind had received an MSS viewer and was observing the imagery on the small screen. With no foliage to clutter the picture now, he could definitely pick out faint fault lines, which corresponded to some of the red vegetation seen earlier. At the intersections of some of these faults, the geologists knew, underground aquifers lay, collecting rainfall from the watershed and carrying it into Lake Champlain. The geologists' objective was to get the imagery to tell them which intersections held the aquifers. The flexibility of the MSS would help them over the next several weeks. They could view the green band, then the red, and then each near-IR band singly . . . then put them together in different combinations . . . enhance one band, reduce another . . . twirl the dials until suddenly a line popped out on the screen . . . was it an aquifer? . . . compare it with the aircraft pictures NASA was supplying . . . next check with ground truth . . . it *was* an aquifer . . . now use the signature to locate another. . . .

The final result was a map showing a network of aquifers. The zoning board could prohibit septic tanks upstream along the aquifers or restrict wells downstream. Unpleasant consequences such as infectious hepatitis and several other major and minor diseases presumably could be avoided.

There was another set of lines discovered on the Vermont imagery, which interested the investigators for reasons completely unrelated to health. This second set of lines represented old river deltas, scattered throughout the Green Mountains. They had been there since the Ice Age, but until ERTS had come along no one had known how to find where they all were.

The Lake Champlain of ancient days had a higher water level than the lake of today, and all the now-dry rivers which fed into it are therefore high up in the mountains. These ancient riverbeds, if they can be found, still contain a supply of sand and gravel. It happens that the city of Burlington needs sand—a scarce commodity in Vermont—for a project converting part of the rocky Champlain shoreline into a beach. It now looks as though the beach will be built with riverbed sand discovered by ERTS.

The Disappearing Beach

After supplying the sand, ERTS will have another assignment at Lake Champlain: to help choose the site to build the beach. Circulation patterns have to be taken into account in establishing a man-made beach if it is not to be washed away in a few years. Beach-goers at lakes and oceans everywhere have had the experience of enjoying a beach area one year and, on returning to it the next year, finding themselves looking around in bewilderment, saying, "Where did it go?"

The beach in such cases had, no doubt, been swallowed up by the water. On an ocean shoreline, it is not unusual to have a hundred feet of beach slip into the sea in a single storm.

Planners of public beaches, as well as the public, find this erosion disturbing. State governments frequently spend large sums of money, in roads and facilities, developing an area, only to have the beach suddenly pick up its sand and move off to become a sand bar perhaps fifty miles away. The flow of sand from one area to another has long been a mystery; and officials like George Armstrong, of California's Navigation and Ocean Development Department, have found it impossible to predict, let alone control. "We have taken periodic air pictures of a beach and tried offshore surveys by boat, but we've only gathered occasional pieces of the sand data we need," said Armstrong.

"It's like trying to count the grains of sand themselves," commented another official.

But now every eighteen days, such officials can receive from ERTS sediment-flow data relative to all the beaches along much of the U.S. ocean coastline. The flow of sediment will be the factor in determining where beaches can be maintained, or where breakwaters must be built in the water to contain the sand. Beach width, wave height, and wind direction are the kind of variables which will affect sand movement, and the eighteen-day ERTS observation cycle has delivered a wealth of this kind of information.

Besides Nature's sabotaging efforts, man himself has contributed substantially to the sand problem recently. For the last several years, marinas have proliferated on the coastlines of oceans and lakes throughout the country. Ocean marinas, especially, trap tremendous amounts of sand along their jetties. If the marina is poorly located, this movement of sand effects a double felony: the sand erodes away from a nearby beach and then proceeds to load up the marina channel. Usually, not even extensive dredging can correct these problems permanently.

ERTS, however, will offer the data base on which to model

the sand movement in computers. The model can then be taken into account before any construction is undertaken, preserving the total coastline and hopefully helping to select easier-to-maintain marina sites.

Dams can be another man-made problem, says Armstrong. "Beaches, even ocean beaches, are sometimes maintained by sand supplied from nearby rivers. Then when dams are built, the sand flow is often virtually eliminated." Sometimes, to everyone's dismay, so is the beach.

The new approach will probably be to keep on file satellite pictures showing the sand flow. Then, when a dam is proposed, sedimentologists can use the pictures to model the effects that reduced sediment would have on the beaches. Considering how avid beach conservationists have been on both Pacific and Atlantic coasts, the new data is likely to be widely used and will supply new dimensions to problems already under lively debate. Once again, ERTS should be able to inject facts into otherwise imprecise arguments.

Controlling Oil Spills

Another kind of pollution is being sensed from the air, not in inland waterways, but off the coasts of the United States. Over a period of weeks in 1970, the Coast Guard experimentally spilled small quantities of oil some forty miles off the coast of Los Angeles. As part of the study, NASA's John Millard and John Arvesen flew over the test area in a remote-sensing twin-engine Cessna to observe the experimental slick. "We saw the slick on the water clearly for a while," Millard remembers.

Then, as he continued his flight, with the sun rising higher in the sky, and the perspective from the Cessna varying, the slick "disappeared."

"We knew right where it was. The Coast Guard vessel was operating right in the oily water, and of course we could see that. But, even knowing exactly where to look, we still couldn't see the slick."

Their sensor was not doing any better than their eyeballs; the slick refused to appear on the recorder in the aircraft for the rest of the day.

"Subsequently, however, we found that by using certain IR and UV wave lengths, the contrast between the oil and water could be reliably distinguished," Millard explained. The position of the sun and the aircraft could be varied, but the slick continued to appear in the gear with impressive regularity when the right wave lengths were sensed.

This meant that the Coast Guard, which often has been frustrated in visually searching for oil violations, could probably improve its performance tremendously by sensing in these wave lengths. Consequently, in 1974 the Coast Guard began conducting experimental operations with UV sensors.

As the Coast Guard and NASA progress together, new techniques and equipment will be developed. In the late 1970s, a situation might occur like this:

A Coast Guard aircraft drones up and down the Pacific coast on patrol. A technician in the after section of the plane stares at a scope—the same scope which has tediously occupied his attention for weeks. He is a bright young man, but bored into a state of mental vegetation. No oil spills have occurred during his patrols, and he drowsily fiddles with his gear. In between yawns, he looks out the window and listens to shore-based commercial radio. An ordinary blip on a radar would have an excellent chance of escaping his notice.

Suddenly, a brilliant purple image flashes onto the screen. The operator happens to be facing another direction, but the brightness of the image catches his attention out of the corner of his eye. A tiny oil slick is being displayed in a brilliant false color, in real time.

As he watches the slick ominously spread on the water, he picks up his hand microphone and transmits, "Slick 250 degrees 15 miles, point Alpha." Almost immediately, a Coast Guard cutter is on its way to the scene. Within minutes of the time the oil began to appear, the offending ship will have been contacted by the Coast Guard. The incident will probably result in a citation, with a warning to halt the oil flow, or it could mean providing early assistance to a ship in trouble and about to initiate a massive spill. The accuracy of the sensing gear and the dramatic color have made it possible to detect the spill early, while it was only a few feet in diameter.

By the 1980s, this same sort of detection can hopefully be accomplished, constantly, over our entire coastline day and night. The equipment? Microwave sensors in satellites.

The Coast Guard technician, instead of being in an aircraft watching a small stretch of coastline, would be in a shore station and could observe all the U.S. shoreline and even the Arctic Ocean. His false-color display would be received from a satellite in real time. Ocean spills that might never had been discovered with 1970 technology would trigger immediate responses in the 1980s. The effort could even be broadened to global scope, with a central station, manned and funded by maritime nations, providing information on violations to "Coast Guards" throughout the world.

International Field Year for the Great Lakes

By the 1980s, satellites will hopefully have collected data defining dozens of pollution problems—concerning oil slicks at sea, lakes being strangled by algae, and streams being bombarded with fertilizers. Satellites will have led to the development of positive guidelines—guidelines which not only prohibit careless pollution but find clean ways of solving the problems of massive modern civilization.

A sign of things to come may have been the International Field Year for the Great Lakes, a project actually begun in 1972 with its final report to be made in 1975. Satellites were only one of several technologies in IFYGL, a Canadian-American multidisciplinary program.

ERTS and NOAA-2 each made significant contributions to IFYGL. ERTS was used to detect lake currents and to monitor other water in the Ontario basin. A composite was made of eight

ERTS pictures, covering the basin. It included the entire watershed draining into the lake. By coordinating ERTS data with ground truth, estimates were made on the volume of water and soil moisture and on the amount of acreage devoted to various land uses in the basin. NOAA-2 was especially useful in determining snow cover.

Along with their extensive data-gathering from buoys, ships, radar, offshore towers, and stream gauges, the satellites helped supply a model of the lake and surrounding area. The goals were to find how the 35 million people living around the Great Lakes (leaping to 70 million by the year 2000) could coexist with the water: drinking it, emptying sewage into it, floating ships in it, powering and cooling plants with it, swimming and fishing in it, and even using it to modify the climate.

In another international Great Lakes program regarding water pollution, the press in 1974 leveled considerable criticism at the U.S. government for failing to live up to water pollution tasks promised in a U.S.-Canadian treaty. Canada was doing her part but the United States wasn't. However, none of this controversy should be confused with IFYGL, a separate and longer range study which indeed was satisfactorily completed. Mapping by ERTS and NOAA-2 of soil moisture and snow cover were a significant part of the final report which is intended to allow sophisticated pollution control in the future.

IFGYL has been just one intriguing water experiment utilizing satellites, and hopefully the results will be encouraging enough to lead to others. The work in *aircraft* sensing of waterways has already embraced significant projects, but satellites have barely been tried so far as water quality is concerned. "Whereas man has begun to scratch the surface of the remote sensing of land by satellite, he has scarcely even put his big toe into the world of water sensing," says one hydrologist. Neither ERTS nor Skylab was equipped with a blue spectral band—a rather major obstacle in attempting to view the ocean. Furthermore, the ERTS orbit is poorly timed for water studies. With its 9:30 arrival time, ERTS is ideally suited over bodies of land to detect the long shadows of man-made buildings, crops, and mountains. But the hour is a poor one for looking at the water. When and if oceanographic satellites become a reality, one of them will presumably have more of a noontime schedule for peering into the depths of a lake or the ocean.

The experiments that have been done up till now have opened several doors, but undoubtedly more will swing open in the future. Just as a biologist's view of a lake from an aircraft, as compared to a boat, is vastly expanded, so will the satellite open the scope much further. The ability to study tens of thousands of bodies of water, instead of one, will shatter the existing limits of knowledge, and ERTS is already previewing the potential of oceanographic satellites.

"We're just barely beginning to recognize what satellite pictures can reveal about water," says Walter Manning of the NOAA research lab in Miami. He comments that each time he looks at a satellite picture under a microscope he finds himself surprised by the details of bridges, ships, and significant water characteristics.

The oceans, rivers, and lakes hold within themselves a large share of the world's mysteries that must be solved if the water is to be maintained clean enough for man's continued and effective use.

14. The Wilderness

In 1912, a Harvard senior named Robert Benchley faced the final examination for his class in international law with a mixture of sporting nonchalance and realistic trepidation. He was abysmally unprepared. The same tendency which in later life would cause him to be tardy in delivering his classic manuscripts to publishers had already characterized his study habits.

The examination required that he analyze the Newfoundland fishing regulations. Undaunted by a complete lack of knowledge of the subject, he vigorously attacked the examination question. Two hours later, he turned in a penetrating analysis of the Newfoundland laws, resourcefully reviewed *from the standpoint of a fish.*

As he read Benchley's paper, the professor reportedly choked with laughter, managing to regain his composure only in time to give Benchley an F, and, incidentally, to keep him from graduating with his class.

If Benchley's essay was unusual, the professor's response was classical. If the same situation had occurred either half a century earlier, in 1862, or later, in 1962, the result would likely have been almost identical.

Today, however, we suspect that the aftermath would have been strikingly different. The professor (a) would probably have seen little humor in the subject, (b) would have given him an A, and (c) would have forwarded the paper to the Friends of the Earth.

Since the beginning of our nation, the American has generally become more and more urban, less concerned with the out-of-doors and the creatures that have been part of it. Then, just a decade ago, along came Rachel Carson with *Silent Spring*, followed by Buckminster Fuller with his concept of Spaceship Earth, and Barry Commoner with his global approach to ecology. A desire to save everything that chirped, grunted, barked, or breathed, and a motivation to hang on to the remote wilderness areas before we unwittingly destroyed them, somehow all developed into a national state of mind.

The fish of Newfoundland waters all of a sudden now *do* concern us. If being fish-oriented was a joke in Benchley's day, it hasn't been one since Miss Carson's arrival on the scene. While only a few people care about fish for the pure sake of fish, a great many are now concerned about Nature's needs, believing they are ultimately our own.

All this has masterfully set the stage for remote sensing. ERTS & Company is arriving in the nick of time. If the trans-Alaskan pipeline were to rupture and dump a few thousand gallons of crude oil into one of the streams along its path, half of the U.S. population would want to know how the spill affected the salmon run. And the fact that such a disaster could be monitored by satellite would mean that persons thousands of miles away indeed might be kept informed that some environmental catastrophe was or wasn't transpiring. The very existence of a satellite might calm waters before they could become troubled. With the whole world watching, a unique quirk of human nature—the desire not to be caught with foot in mouth—should become a factor; environmentalists would think twice before making irresponsible accusations, and pipeline authorities would likely become pristine ecologists overnight.

The Pipeline Disaster Patrol

The ability to sense oil spills is discussed in other chapters, so we need not dwell on those sensing systems here. What we might mention just now is the locking of the barn door before the horse can be stolen: the Alaska pipeline is likely to be the subject of considerable monitoring, all in a prudent effort to avoid disasters before they can arise.

A sensing satellite could provide data as to the status of both the pipeline and its "thawbulb," the soil which surrounds the thousand-mile tube of steel. The super-hot oil line would be a constant threat—it could melt enough ice in the thawbulb to create floods or, worse, melt saturated permafrost and cause massive brown flows of mud to bubble across the tundra. A satellite could detect these problem conditions shortly after melting began, making it possible to keep the situation from developing into a costly spectacular. Its task would be to alert emergency crews both to the possibilities of such lesser calamities as floods and to the likelihood of the ultimate disaster, a pipeline rupture.

An environmental study by the U. S. Geological Survey anticipated substantial disruption of the thawbulb during the first five years of the pipe. During this time soil is expected to settle, conceivably to the point that it could cause the steel pipe to flex excessively and crack open. Well before this stage would ever be reached, however, the alert satellite will hopefully have sensed the stress and signaled to its masters that big trouble was on the way.

In 1970, the USGS contributed substantially to the environmental-impact paper prepared by the Department of the Interior on the pipeline. In the paper, USGS geologists reviewed the proposed pipeline route and outlined potential hazards with an insight which impressed even Sierra Club hierarchy. The USGS also outlined safeguards which would be necessary if the pipeline were approved.

The same agency's geologists were later assigned, in large measure, the responsibility for keeping the line safe, thereby testing their own recommended safeguards. Their incentive to keep the line safe thus encompasses environmental concern, professional pride, and career survival, a combination well calculated to achieve results. Their interest in satellite data is sure to remain keen.

From the earliest ERTS pictures, the USGS studied Alaskan images carefully. Knowing that oil exploration during the 1940s had left the terrain scarred in various remote areas, geologists examined ERTS pictures to see how the landscape appeared thirty years later. Their judgment, based on the pictures, was that the terrain seemed to have recovered from the destruction of the 1940s, as mentioned in Chapter 5. Geological teams visited selected areas in 1973 to ground-truth the ERTS findings. "The field work substantiated the ERTS information," said William Fischer of the USGS. "It showed us that the environment in Alaska would not be permanently devastated by pipeline construction handled with reasonable constraint." The information was provided to Congress prior to the final decision in late 1973 to approve the pipeline.

Vocal opposition to the USGS in Alaska has frequently come from the Sierra Club, which spoke out against the pipeline from the beginning. The club throughout the early 1970s published long lists of reasons for their opposition. Their objections were varied, concerning potential damage to vegetation and permafrost, siltation (as well as possible oil pollution) in streams, and earthquakes and landslides. Although the Department of the Interior considered many of the same potential disasters in the environmental report, their final conclusion was essentially the opposite—that is, that the pipeline could, with reasonable safety, be built. With the awareness of hazards existing on both sides, however, it should hardly come as a surprise that some Sierra Club members now consider it their duty to guard the wilderness during the period of construction and early use of the line.

Several of the club's early objections to the pipeline concerned wildlife in Alaska, especially the caribou and salmon. Since a large proportion of the pipeline was expected to lie on the ground, environmentalists had predicted that the migrating caribou would be halted by this "fence" across the tundra and would follow the pipeline south all the way to Fairbanks.* The Sierra Club felt that our 50,000 Eskimos, whose traditional livelihood had made them dependent on these animals and fish, de-

* Later plans were made to have much of the line elevated by trestles or buried underground, depending upon the terrain.

served some consideration, notwithstanding the nation's strong reliance on petroleum. These and many other factors have caused at least one of the club's Alaskan authorities to show an interest in reviewing ERTS data.

The data is conveniently located just thirty miles south of Sierra Club headquarters in San Francisco. There, in Menlo Park, California, is one of the libraries containing a complete ERTS catalogue and many other satellite pictures; it is open to the public. There is no doubt that some of the club members' objections to the pipeline can be evaluated by a continuing study of ERTS data. As mentioned before, pipeline ruptures and changes affecting broad areas will be apparent in ERTS images. At one time, it was felt that the migrating caribou themselves would appear on the ERTS image. It was thought that the immense mass of animals, moving across Alaskan mountains and valleys, would be large enough a subject to be well within ERTS resolution. They were expected to be our classic pickle-barrel image, appearing on ERTS as one mighty blob. In reviewing ERTS data thus far, however, this does not appear to be the case. Whether the Sierra Club will be able to study the migration indirectly, by observing surrogate images—such as the alteration of vegetation and snow caused by the animals—is not yet known. But it is certain that the animals' impact on the present environment, on the construction of the pipeline itself, and on the economy of the Eskimo are all matters which will interest various groups.

The idea of actually seeing herds of animals on satellite pictures is a novelty, and obviously not the means by which most wildlife-sensing work will be accomplished. The forecasting of populations of ducks and geese in North America offers a more conventional example. For many years, the Bureau of Sport Fishing and Wildlife has used aerial photography to measure standing water and vegetation in areas known to support waterfowl. By measuring the water and feeding grounds available to the birds each year, the bureau derives estimates of bird population and then establishes limits for hunters accordingly. By controlling the hunting seasons, the populations can be increased or decreased as desired for the over-all benefit of the sustained existence of waterfowl. In 1973, BSFW commenced experimenting with ERTS pictures in an effort to determine if satellite data could become a substitute for aerial pictures. The expense of flying aircraft to map the water and vegetation in vast areas of North America was one which satellites might help reduce in the future. With this in mind, ERTS pictures of the north-central United States and Canada in May and July 1973 were studied by BSFW's Harvey K. Nelson in Jamestown, North Dakota.

Nelson said that he found ERTS effective in mapping larger ponds and areas of standing water. When the satellite pictures were interpreted, Nelson found that they, like the aerial pictures, indicated that it would be a bad year for the mallard duck and certain other migratory birds. Hunting seasons and limits were reduced accordingly in 1973, says Nelson.

Nelson has now applied to NASA to make an ERTS-2 study in which he would compare the traditional aerial survey methods with a new system involving satellite pictures backed up by scattered airplane photographs and ground truth—including a financial comparison of the two systems. "I think we might find that we can obtain better information at less cost with satellites," said Nelson. Furthermore, the satellite information gathered in waterfowl studies would be ideally suited for research in water quality.

The Balance of Life in Nature

Not only for waterfowl, but for all animal life, satellites seem capable of showing the vegetative condition of the wilderness, helping wildlife managers to decide where animals and birds can be supported. One outdoorsman interested in the balance of wildlife is David Carneggie, the University of California rangeland investigator introduced in Chapter 8. On July 26, 1972, just three days after the ERTS-1 launch, Carneggie was on his vacation, camping in the High Sierras. He noticed that the grass was still lush, and he saw a herd of deer grazing. He was in a prime rangeland area for deer, although cattle were also allowed there during portions of the year. As it happened, all of this took place on the day that ERTS was sensing that very area for the first time.

Back in his office, Carneggie eventually saw the imagery for the green grass where the herd of deer had been grazing. The image of this grazing area was, of course, just as suitable for estimating yields as the rangeland pictures we saw Carneggie routinely studying for the cattleman in Chapter 8. Eventually, he conjectured, the Bureau of Land Management would be using ERTS data to determine just where a rangeland should have cattle removed to prevent overgrazing and to support the deer population. And then when satellite data would reveal that rangeland had become too scarce to support the wildlife (because of drought, overgrazing, or overpopulation), the Bureau of Sport Fishing and Wildlife would have the option of expanding the hunting season.

The balance of cattle, people, and wildlife, Carneggie reflected, was not an easy one to strike—a fact recognized not only by himself but by conservationists throughout the world. Richard H. V. Bell, of the huge Serengeti game preserve in Africa, had recently put Carneggie's thoughts into words: "It is up to man the scientist . . . to suggest integrated patterns of land use, taking into account the interest of man the pastoralist, man the cultivator, and man the tourist."[1]

ERTS—Watchdog for Environmentalists

There are many remote areas in the United States which man the tourist only occasionally reaches or sees. This fact has dismayed those tourist-environmentalists who are convinced that the U. S. Forest Service and private timber companies are in a conspiracy to eliminate the wilderness. Accusations about vast forests being cut and not replanted are common. Whether or not there is truth to these allegations has occasionally been burdensome to resolve, because of the remoteness of the lands concerned. With satellite imagery, however, this should cease to be a problem, since government policy dictates that the images be in the "public domain," a phrase which in actual practice may have different significance for different groups. For a well-financed environmental organization, the phrase should mean that pictures be purchased on a regular basis and that a highly skilled photo interpreter be hired to view them. In fact, groups with the wherewithal to have their own equipment and talent are generally now going to automated electronic processes. Universities are generally in this category, and we mentioned earlier how an oil company had gone even further by purchasing global tapes and storing data of the entire world in a computer.

As for the individual person desiring inexpensive information of a particular area, he has the option of purchasing random pictures for his own inspection. I first tested this system myself in 1973. I had seen an ERTS picture (the garbage-ship wake off Long Island) of which I wanted a copy to incorporate into this book. I picked up the phone and dialed (605) 339-2270, the number of the data center at Sioux Falls, South Dakota, where the Department of the Interior acts as a sort of official central librarian for satellite pictures. A voice identifying its owner as one Mike Delaney, responded, *sounding anxious to help*. I identified the picture I wanted by number (and designated that I wanted the red band), and Delaney accepted my order for immediate processing. The cost was $1.75, and I received the picture within a week. For more thorough research, one can visit any of the some forty browse files located throughout the country†—to find pictures either for ordering or for on-the-spot viewing. A word of caution to would-be amateurs: Don't attempt to draw conclusions about ERTS pictures without help from a qualified photo interpreter.

With the Department of the Interior's efficiency in providing pictures, I rather suspect that there will probably be a number of situations occurring in the future where environmentally

† A national list of browse libraries is available from EROS, Department of the Interior, Washington, D.C. 20242.

oriented individuals will take it upon themselves to challenge various government entities, hopefully after guidance from a good interpreter. I can visualize confrontations becoming common, perhaps occurring rather like this:

An environmentalist obtains several pictures of a remote forest area and pays a visit to a Forest Service office. "It looks to me like someone is clear-cutting timber stands," charges the environmentalist, pointing an accusing finger at a large area in pictures where a distinct absence of vegetation is apparent.

"Perfectly all right," the Forest Service might conceivably answer. "That's a recommended procedure for the Douglas fir in that locale."

"And what happens now?" the challenger persists.

"We'll see that it's replanted for sustained yield," would be a sensible official response.

Subsequent satellite pictures can later be examined by the environmentalist to verify that the planting indeed does occur. Conservationists who understand that some cutting is desirable for the forest's vigor, and who know that some access roads to remote forests are necessary to perform such tasks, will make good use of the synoptic view. Uninformed critics may cause official distress from time to time, and, just as surely, irresponsible officials or logging companies will suffer under the spotlight of this information in the public domain. But while sensing data offers the public a chance to look distrustingly over the shoulders of our federal agencies, the agencies themselves will surely use the data considerably more than anyone else. Thus far the Forest Service would seem to have emerged as an enlightened user of remote sensing for accomplishing its varied responsibilities.

Broad Tasks of Forest Service

To preserve the wilderness, the forester must maintain wildlife, prohibit clear-cutting of timber in some situations and encourage it in others, keep trees and vegetation healthy, and control fires and floods. The forester realizes that man is wasteful when he doesn't replenish woodlands with new growth, tree for tree—and yet almost as wasteful when he doesn't utilize every renewable board foot of timber the forest offers. The people of the United States started two and a third million houses in 1972 and, largely for that purpose, cut more than 40 billion board feet of timber for the first time. The forests were straining to meet the demands—and not achieving it; we had to import 10 per cent of the softwood we put to use. Regardless of what actual or political factors caused this timber shortage (and the spiral in costs which had become acute by 1973), any additional mismanagement of forests will surely increase it. Fire, disease, and insects are all big factors affecting the supply and cost of lumber.

With its Multiple-Use-Sustained-Yield Act of 1960, Congress gave the Forest Service the responsibility of managing the nation's woodlands along multiple-use guidelines. Congress seemed to agree with the concept that man was "a pastoralist . . . cultivator . . . tourist." All forests, saith the Act, shall simultaneously grow timber, support grazing wildlife or cattle, and provide water and recreation.

To observe all these concepts would require, to say the least, a rather broad overview of the forest land of the United States. And yet the Forest Service has fewer then 1,000 rangers for managing 850 million acres of generally remote land. The logical device, of course, for getting to know the territory is none other than the earth-orbiting satellite.

We should point out, however, that for thirty years before the first satellite was launched, the Forest Service Survey was firmly established in the business of inventorying its entire timber resources, covering all portions of the United States every ten years. The result has been an impressive map and data bank indicating both the acreage and the board feet timber potential of all major tree species. The Survey has been conducted continuously, so that at any given time a small part of the map has always been completely current, while some of it was as much as ten years old.

Synoptic View of the Forest Tested

Since March 1969, it has been apparent that satellite pictures might revolutionize the Forest Survey. On that occasion, a set of 80-millimeter cameras mounted on the Apollo 9 capsule snapped a four-channel picture of a 7,600-square-mile forest area in Arkansas, Louisiana, and Mississippi.

The single satellite photo covered an area which would have required no less than 35,000 of the small, high-detail photographs (1/20,000 scale) the Survey customarily used.[2] However, if this seems to imply that satellite pictures could immediately render aerial photography outmoded, let us hasten to say that in the case of forest surveys they would not. The idea of a satellite taking a decade's worth of aerial pictures with a few fast clicks of the shutter in a single afternoon is exciting but unfortunately not practical in the case of trees. Satellite resolution cannot yet make the subtle distinctions among species of foliage of which aircraft photos are capable.

All that foresters hoped for in 1969 was that the Apollo picture would prove adequate for the first stage in survey work—distinguishing forest from nonforest; revealing those areas which had been switched from forest to croplands since the previous census; differentiating between the complex forests which would require a closer look by airplane and marginal ones which would not. The Apollo picture was studied with this in mind, after which airplanes and ground surveyors were sent in for detailed sampling in key areas. Their results were interpolated and pyramided into the Apollo picture where detail was needed.

The result was a completely satisfactory forest-vegetation map. Instead of a 35,000-picture mosaic, with all its inherent tonal variations, it was one uniform picture with aircraft close-ups inserted where necessary. Only 2,000 aircraft pictures were required, and thousands of tedious and expensive miles of flight had been avoided. The expectations of the Forest Survey had been dramatically exceeded.

Simple arithmetic revealed that in operational use it would take only 80 such satellite pictures *to cover all the forest land in the South.* And if satellite pictures were provided nationally, followed by the usual air and ground work, the ten-year survey could probably be quite handily cut to five.

Unfortunately, the next opportunity to utilize a satellite was not until 1972. By then, the Forest Survey was prepared to use ERTS not only to map forested areas but to obtain more specific information; some 850 check points requiring forest classification data, all in the state of Georgia, were assigned to ERTS. ERTS was also expected to supply a current environmental picture: define the cutting status, identify burnt-out areas, locate power lines, and outline the urban encroachment of forests. Instead of the single shot of 1969, the repetitive pictures of ERTS hopefully would detect seasonal variations and conceivably unearth a few unexpected discoveries.

It did both. Among the expected achievements was an effective vegetation map. By combining January and February ERTS pictures, foresters could detect the differences between leafless hardwood trees and bushy pines, forming a forest mosaic for the entire area. Among the unexpected items was the development of a means of mapping the kudzu, a regional vine which overgrows trees (as well as large areas of farmland) and eventually kills them. The kudzu vine identified itself unmistakably by appearing bright red in October IR images and dead and gray in April; it was the only vegetation to follow this pattern. The locations of the kudzu were reported to the regional Forest Service authorities and county agriculture agents, who can now pinpoint their problem. Kudzu, which would tend to retake an area rapidly if not thoroughly eradicated, can be destroyed only by a massive attack with over-all direction. ERTS is offering the means.

While ERTS was mapping forest features throughout Georgia in the summer of 1973, a laser device carried by aircraft was also passing over southern forests. The "laser profiler" was flown over Sam Houston National Forest in Texas in a successful preliminary attempt to measure the heights of trees. The profiler has the potential of becoming a satellite (as well as aircraft) sensor, and in such a role would be an ideal teammate to scanners or cameras in taking forest inventories.

The laser profiler has to be viewed as highly experimental at this time, but remote-sensing experiments obviously are progressing rapidly in forestry, ERTS and Skylab having reached their present levels in a relatively short span of time.

Satellites as Potential Fire Alarms

In 1972, of course, the success of ERTS was still only a hope, not a certainty. In late August, just a month after the launch, I was visiting two of the investigators who were then still looking forward to working on the Georgia project; they were Forest Service research scientists Robert Aldrich and Robert Heller. It happened that I paid my visit on the day when they had just received their first ERTS imagery. It was an MSS composite view of Alaska taken three or four weeks before. Both Aldrich and Heller were anxious to begin receiving images of their own project areas, but it was to be several more weeks before ERTS data would flow smoothly from satellite to receiving station to individual scientist. The chore of cataloguing imagery for the public browse files, as well as for investigators, was at that early stage bogged down. The Department of the Interior, which was handling the data, was placating such investigators as Aldrich and Heller by sending them occasional random pictures simply to satisfy their curiosity about ERTS.

Heller and Aldrich were consoled by the fact that this first ERTS picture sent to them was, at least, a forestry scene—a 115-mile square taken from an ERTS sweep through the northern part of Alaska. Heller looked at the upper-left-hand corner of the picture with his naked eye. "Look at that burnt-out area," he said. There was a black spot in the midst of red vegetation. "That looks recent."

Aldrich pointed to the right side of the picture. "And there's more on the way." They could see bluish plumes of smoke streaking westward from what was obviously a forest fire in progress.

Aldrich mentally measured it. "About twenty miles long already. That's quite a fire."

That it was. It seemed to be veritably jumping out of the picture demanding that a fire-alarm bell be rung—a procedure, incidentally, that some sensing enthusiasts have suggested might be accomplished by satellite.

The Alaska fire was much larger than the minimum qualifying size which Craig Chandler of the Forest Service had designated for "high intensity" (HI) blazes.[8] In 1971, Chandler had thoroughly reviewed prevailing concepts of using satellites in fire detection, in response to the growing belief that satellites could virtually take over fire-sighting control. The use of far-IR to detect heat waves was the tool being anticipated. Chandler's analysis, however, indicated that if a satellite sensor were to sound a fire alarm for every blaze it could detect, fire fighters would be charging in on every campfire in the wilderness. Satellites, he concluded, were not practical as primary fire-alarm boxes, and he pointed out that the need for satellites lay in detecting the HI fires that overwhelm ordinary fire-fighting forces.

Under present operational methods, no one can tell that a fire ranks as high intensity until forces arrive on the scene and find themselves hopelessly outgunned by the blaze. This delay in recognizing HI fires results in considerable loss. While only 3 to 8 per cent of total forest fires rank as high intensity, Chandler says, they do fully 90 per cent of the damage—hundreds of millions of dollars' worth of timber, homes, and wildlife each year, plus nearly $70 million in fire-fighting expense.

Without much difficulty, however, an IR satellite could be equipped to respond *only* to HI fires, sounding real-time alarms. The alarms could immediately be fed into a predetermined battle plan on the ground.

An important key to fighting fires is obviously the need to act early. A blaze which might be child's play to combat in its second day could be a raging fury in its third. How much of the HI losses could be avoided by a satellite system is pure speculation at this point. But the annual loss (of timber and property) from HI fires runs approximately double the budget for the entire earth observations program, covering not merely forestry but all remote sensing!‡

Insect Damage Discovered by Satellite

The challenges of caring for the wilderness areas are further brought into perspective when we note that fires, costly as they may be, are ranked as only the number-three problem in timber loss. While fires are still the most spectacular enemy of the forest, they damage fewer trees than such silent enemies as *Dendroctonus frontalis hopk.*

In 1963 and 1964, *Dendroctonus*, an unobtrusive six-legged creature just an eighth of an inch in length, invaded the timberlands of Central America and succeeded in destroying 60 per cent of the Honduran pines almost before it was noticed in the dense growth. Then, after two years of conquest, it lost interest in Honduras and sought greener forests elsewhere.

The *modus operandi* of this international beetle is to creep under the bark of the pine, carrying a fungus along with it. The results are disturbing: certain death for the blighted tree, plus the likelihood of an exponential invasion of surrounding trees. Thus, 5 blighted trees in 1976 could become 25 in 1977 and 125 in 1978.

Dendroctonus has a tendency to appear at random points, scattering its family throughout entire forests, rather than concentrating on a single area. The task of locating a band of these beetles who are spreading themselves over millions of acres of remote timber has proved to be more than Herculean, at least until now.

In 1970, the beetle became active in the United States, seeking out virtually all the varieties of pine extending from Appalachia to the Pacific Northwest, and launching minor attacks

‡ The NASA applications budget for 1975 is approximately $177 million.

on Douglas fir and some other tree species. By 1972, it had killed more than a million lodgepole pines in Yellowstone National Park and the scenic Jackson Hole country alone.

The Black Hills of South Dakota proved to be another of its favorite haunts. For that reason, both ERTS and Skylab were called upon in 1972 and 1973 to locate clumps of the ailing trees throughout the Hills, with Robert Heller again an assigned investigator.

ERTS images offered what seemed like a simple but promising visual means of detection; ERTS interpreters could look for yellowing trees in the green forests. In an operational system, such pictures might be reviewed by the Forest Service every eighteen days in an effort to locate, then quickly cut and destroy the trees—and, hopefully, *Dendroctonus* along with them. But this kind of sophisticated reaction was not planned in Heller's work, which was of course experimental. It was to be a pilot study, gaining experience with this one beetle which could be used later to detect all varieties of the two leading enemies of the forest: insects and plant diseases.

However, the early results from ERTS made the outlook for the experiment extremely discouraging. Extensive cloud cover over South Dakota in late 1972, followed by the snow cover of early 1973, prevented any detection. Heller was briefly optimistic in the spring of 1973, when ERTS pictures began to provide clear views of the trees. Unfortunately, said Heller after reviewing the pictures, "the yellow of the infested areas doesn't have enough contrast to be distinguished from healthy green trees." Visual use of ERTS imagery seemed to be ruled out.

Meanwhile, in June 1973, Skylab astronauts took a photograph of the Black Hills region, revealing high contrast of infested areas, and Heller felt that its potential was outstanding. Since there had been no advance notice of the astronauts' picture, no ground truth had been obtained. However, areas which Heller believed to have infested trees "showed up clearly, sometimes with a resolution of thirty feet," he said. A resolution of that accuracy meant that a single tree could sometimes be distinguished!

The June Skylab pass over the Black Hills also offered an exciting possibility completely apart from photography—far-IR sensing by the Multi-Spectral Scanner. The thermal detection ability of the MSS lent itself to detecting *Dendroctonus* because the physiological reaction of trees to the beetle begins before the tree changes color. As the beetle's fungus spreads under the bark, the tree loses moisture and the pine needles get warmer; in short, a sick pine tree runs a fever, sometimes as high as nine degrees Fahrenheit above normal. Skylab's MSS has already been used in aircraft in similar experiments and proved itself capable of detecting the tree-temperature variation 25 per cent of the time *before the pine needles turned yellow.* This "previsual" detection ability meant that Skylab might be able to tell us if a pine tree was ailing—even though a forest ranger standing with his hand on the tree trunk might never notice it!

However, Heller and his associates have not yet succeeded in using the MSS tape to detect those trees with higher heat. Therefore, as of late 1974, neither ERTS nor Skylab techniques have proved themselves in the Black Hills. Both, however, are being pursued optimistically. In the case of ERTS, where we have already mentioned that the problem was one of green not being distinguishable from yellow, the examination is now being automated; experiments with tapes are being conducted to find out if electronic comparisons can detect the yellow trees which evaded the human eyeball. This degree of improved interpretation is one we would expect to progress rapidly, with such applications discussed earlier—corn blight and strip-mine reclamation—being striking examples of the uses to which the technology will be turned.

The problem with the Skylab MSS, as with ERTS pictures, also involves skill in interpreting tapes. The IR tapes are now being enhanced by Heller's colleagues in an attempt to make the high temperatures of infested trees distinguishable. The tasks attempted with Skylab, singling out individual trees and taking their temperatures, could conceivably be performed in the near future, although probably not as soon as basic progress is made with ERTS and its automated color coding. But thermal detection will eventually provide us with a skill that should quickly revolutionize the monitoring of the health of forests.

The sensing of stricken trees is just one of half a dozen tasks in the forest which might become part of operational satellite patrols. By 1980, when the Space Shuttle is orbiting, the sensing of forests will hopefully be an advanced, multipurpose art. The Shuttle's lower (100-mile) altitude may allow it to single out individual trees even more dependably than Skylab, although the Shuttle will not have the blanketing ability of ERTS.

Operational Satellites for the Wilderness?

A combination of these two assets—high resolution and comprehensive coverage—is what must be ultimately achieved if satellites are to effectively sense forest lands by the millions of acres. And the requirements of the forest are heavy enough so that they might well require a permanent satellite, one that could be launched exclusively to monitor forests and a few other resources. A study for the University of California, conducted by Richard C. Wilson, former director of the U. S. Forest Survey, proposed this very thing. He advocated that by the 1980s the Forest Service should finance a fourth of one full-time satellite. He had in mind increasing land and timber classification to five-year frequency; inventorying wildlife habitats yearly; monitoring recreational resources frequently; monitoring disease and forest weather damage twice a year; monitoring snow and water conditions daily during flood conditions; forecasting potential fire weather daily during fire season; and monitoring air pollution every other day following fires.[4]

Wilson's projections relate only to the United States. Actually, however, the forests of the tropics contain the largest share of the world's hardwood treasures. ERTS is providing the first, very general, worldwide inventory of these forests, which will simultaneously be of international interest to those interested in using the trees and those oriented toward conserving them.

The full-time monitoring of forests on a global basis could logically be complemented by the sensing of other remote phenomena we have discussed. Approximately 27 per cent of the earth's land surface is forest land, an estimated 26 per cent is permafrost,[5] and 20 per cent is thought to be rangeland (although only a satellite knows for sure, and therefore inventorying these acreages would be among the first assignments an operational satellite would be asked to perform). However, assuming for the moment that these estimates have some accuracy, we can see that a satellite monitoring nothing but forests, ranges, and permafrost would be sensing roughly three fourths of the world's land, and so would presumably have enough to keep itself reasonably busy, full-time.

The value of rangeland monitoring was discussed fully in Chapter 8, but we might say a word or two more about the importance of permafrost monitoring. We have already discussed it in relation to the Alaska oil pipeline, because of that project's relevance to the 1970s, but the pipe thawbulb is, of course, only one tiny portion of the world's frozen ground. Awareness of the depth and distribution of all the world's permafrost is crucial for many kinds of development in cold zones.

The planning of highway construction is one good example in which knowledge of permafrost and soil conditions is critical. Ernest Lathram of the USGS points out that vegetation in ERTS imagery offers the key to Alaskan soil types. Fine and coarse soil can be distinguished by their vegetation. "Fine soil is sensitive," he says. "When ice melts and contracts, fine soil can sink to what would be a disastrous level for road construction." An understanding of thesc factors requires some degree of *repetitive* study; wilderness areas (much like oceans) require man's constant surveillance.

In our discussion of the pipeline, we viewed ERTS much as though it were a police stakeout guarding a bank after a tipster had indicated a robbery to be in the offing. If a metropolitan policeman were actually in this situation, sitting in his car outside the bank, other robberies would probably be occurring throughout the town, in places both predictable and unlikely, with no police around. A police force obviously cannot spread itself so thin as to guard the entire city.

Yet this is precisely the advantage of satellites. They *can* be everywhere at once, providing a repetitive picture of remote areas. The pipeline rupture is just one of many potential eco-hazards which need to be monitored in the wilderness; the list of destructive phenomena is a long one, ranging from constant minor problems, such as soil erosion, to occasional disasters, such as Alaska's galloping glaciers. These gargantuan chunks of ice are capable of flowing hundreds of feet a week and colliding with civilization full force. They may remain motionless for years, then begin to move suddenly and dangerously. The movement can be detected and measured only by a repetitive overview, and once again, satellites offer the means to protect man's resources—not only from his own frequent ineptitude but also from Mother Nature's constant mischievousness.

The wilderness, with its remoteness and inaccessibility to the human eyeball, obviously needs tending after. The great forests and other growing things are part of the limited supplies on Spaceship Earth—supplies which we must not abuse and yet must utilize to their fullest to satisfy our present and future needs. Unlike fuels and ores, the renewable resources allow us not only to eat our cake but, if we truly keep an eye on things, to allow posterity to have it too. And while this is a challenge which involves considerably more than satellites, they are the first indispensable step toward total conservation.

Part III

THE FUTURE

15. The Future

In 1895, a twenty-one-year-old Irish-Italian lad named Guglielmo Marconi emerged from several years of work in his father's vegetable garden near Bologna and revealed to the world his wireless telegraph for electromagnetic radiation. His discovery came only a decade after a Scotsman, James Clerk Maxwell, had proved by mathematical analysis that electromagnetic waves existed and traveled at the speed of light.

Marconi's discovery led to a variety of early applications for the wireless, none going much beyond the narrow confines of one-to-one communication. In 1908, a manufacturer who was selling radio transmitters to the U. S. Navy broadcast the efforts of a musical group from the Eiffel Tower, apparently in a spirit of public relations gimmickry. Similarly, in 1910, Enrico Caruso and the Metropolitan Opera were transmitted "live" from New York for the benefit of nearby ships and, more specifically, for a group of reporters clustered around a receiver in New York. What now seems an obvious conclusion—that radio was ready to expand from the headset fraternity to mass audiences—was apparently not even considered. The wireless was not for entertainment. It was for important messages.

In 1912, another twenty-one-year-old confirmed for the world its belief that radio was indeed a tool for vital communications. Sitting at a radio receiver in New York, the youth received the dots and dashes that spelled out, "The S.S. *Titanic* ran in to iceberg. Sinking fast." The incredible news ripped through New York like wildfire. For three days and nights, the youth sat glued to the receiver and exchanged messages with the rescue ship

Olympic, 1,400 miles away, while crowds clustered outside the building listening for names of survivors.

It was not until ten years after the *Titanic* episode that the wireless operator, since advanced to the role of young executive, was heard from again; it was then that he proposed to a wireless company the unique idea of selling "radio music boxes" to households and broadcasting music to them on a commercial scale. Response from the company management was less than overwhelming, but the young man was not inclined to give up early—for this youth who had spent three days and nights receiving dots and dashes was no less than David Sarnoff, who more than any other single individual was to build RCA and NBC, and develop the entire radio-television industry in the United States.

It seems safe to say that neither Marconi nor any of the other fathers of radio at the turn of the century foresaw its mass application. Certainly none of them visualized Radio Detection and Ranging (acronym: RADAR). And James Clerk Maxwell probably did not, even in his wildest dreams about electromagnetic radiation, have any thoughts about remote sensing as it is known today (or, if he did, he prudently kept them to himself).

Nor did Marconi and the other early utilizers of active electromagnetic radiation have any reason to exchange ideas with the developers of cameras, and surely neither of these two separate scientific groups realized that some of their efforts would converge in the mid-twentieth century in the form of remote sensing! Like the early developers of the airplane and the automobile, they visualized only a tiny portion of the pursuits to which their inventions would be applied. Even these men of extraordinary vision managed to see only a thin band in the "spectrum of life" about them.

Whither Goest the Satellite?

By glancing backward and taking note of man's inherent myopia, we may better appreciate the unpredictability of the future. Into what earthly pursuits will spacecraft be leading us by the year 2000? Will their primary task be one of exploring the solar system, or even the universe? Will they identify planets bountiful enough to merit close examination? (If earth is a small oasis in space, as astronaut Frank Borman has described it, is Jupiter perhaps a larger one?) Or will spacecraft prove to be more significant when they remain close to home in earth orbit, serving as weightless laboratories for research in cancer or other disease? Will satellites become orbital industrial parks, cosmic outposts of earthly manufacturing, developing products which could improve the earthling's life in a multitude of applications? Or, if we judge from the histories of radios and airplanes, will the primary mission of satellites be something as yet not even conceived?

These are all very credible possibilities, concedes Anthony J. Calio, Director of Science and Applications at the Johnson Space Center, but there is ample room yet available for imaginative concepts within the young field of remotely sensing the earth, and it is down this still lightly trodden path that Calio's own conjecture tends to roam. He suspects that sensing itself will undergo the metamorphosis of the wireless, the airplane, and the horseless carriage, and he anticipates that current sensing achievements are quite likely the precursors of revolutionary applications which have not yet even flickered in man's less responsible dreams.

Men are already wondering when sensing systems will leap beyond their present narrow spectral confines to absorb new expanses of information and seek new combinations of knowledge. Will science find effective ways of integrating information from microwaves with geophysical data—and instead of depending on sensing the earth's surface to estimate the likelihood of petroleum deposits, will sensors instead penetrate the crust of the earth to detect subterranean pools of oil? While Calio may not ponder on that specific concept, he certainly foresees an era not far off when the spectral sensing band will include wave lengths unmastered today; an era when even the data being collected in the 1970s will be interpreted with a new precision that will open fresh areas of knowledge; an era when computers will be stored aboard satellites, critically selecting the information which will be retained and analyzed with efficiency many times that realized today. All of these technological advancements will open doors to unknown opportunities. It was late one evening in 1972, while excogitating the possibilities ahead of us, that Calio prophetically summarized it by stating quite simply, "The big payoff is yet to come."

The More Immediate Future of the Late 1970s

We can be somewhat more specific about sensors in the years immediately ahead. Skylabs 1, 2, and 3 have completed their missions but have left considerable material yet to be analyzed. ERTS-1 is continuing to provide useful information. The 1975 Apollo-Soyuz is at present not scheduled to include earth observations projects—a reasonable decision in view of the project's brief (ten-day) duration. ERTS-2 is developing as a "quasi-operational" satellite with a considerable amount of the information it will collect being scheduled for immediate use.

Looking past ERTS-2 (and having to disregard such unresolved proposals as SEASAT), the next great sensing opportunity looming ahead is the Space Shuttle. The Shuttle will probably be first flown in 1977, with orbital tests in 1979 and no operational sensing until the 1980s. The Shuttle will be a departure from all space activity to date, even including Skylab. Whereas Skylab is an orbiting laboratory to which three conventional spaceships traveled and then returned to earth by means of splashdown, the re-usable Shuttle spacecraft is a veritable turtle carrying its house on its back. The Shuttle spacecraft will be launched from earth, become its own space lab, obediently return to land on earth like an aircraft, only then to find itself being shot off again and again. It will spend only a few days or weeks in space each trip. It may dock and orbit there with a work station, but except for mini-satellites which it may put into orbit, all the remote-sensing equipment will be in the Shuttle itself.

Today in the pre-Shuttle era of one-time-use spacecraft, it costs us several thousand dollars to place a pound of equipment into orbit. With the Shuttle, the cost will shrink to perhaps a hundred dollars. The sensing equipment can therefore be more extensive and varied and hence cover a broader frequency range. Radar gear, which has been previously unusable because of its weight, will be possible in the Shuttle, and active microwave devices to which we have frequently and longingly referred can be a reality.

"ERTS was a great moment in history," says Calio, "but like most great 'firsts,' it will be followed by a period of sophistication. The fast turn-around time of the Shuttle *will make remote-sensing experimentation perhaps a hundred times more productive!* Manned orbiters will be shuttling to and from the work station, giving us the flexibility to shift emphasis, to choose from a whole variety of experimental options, including both equipment and procedures. We can change wave bands, or replace sensing gear, or switch to new variations in the experiments themselves.

"With ERTS," says Calio, "a scientist has limited flexibility —he's like an artist who starts one sketch and has to stick with it, whether he likes it or not, because he has only one piece of paper. If a scientist is, say, trying to monitor a watershed, he may work on an ERTS investigation for two years and perhaps not have it yield the information he wanted."

With the Shuttle, Calio says, a scientist can be like an artist who tries several sketches and selects the best one. He might perform extensive exploratory work on a January Shuttle, examining several courses of action for the watershed project. Then when the spacecraft returned to earth, he could examine the data. He might rule out several wave bands and settle on a few good ones. Then he could get his experiment back aboard the spacecraft in March, with a combination that had a high probability of success. In a few short months, the Shuttle would have ruled out several concepts, and then finally proved one that could be immediately available to man.

"Once *that* is accomplished," Calio summarizes, "the scientist is capable of equipping an operational satellite to monitor the water project regularly and dependably."

In order for this kind of operational program to become effective, however, Calio says that considerable efforts must be made in handling the data after it is collected. "Our task is to develop not only a sensor, but a total system. The data cannot be voluminous and cumbersome," he says. "A big challenge exists in the 1970s to weed out unnecessary data, to learn to interpret data properly, and finally to put it into a useful operational form."

Spy in the Sky

Technological challenges are not the only obstacles faced by remote sensing; public opinion is another. There are, for example, many persons who look on the sensor as a threat to individual privacy. They see NASA sensors as civilian versions of the military's Spies in the Sky; just as the great powers keep each other under military satellite surveillance, so could NASA's sensors inopportunely peer down on each of us in our personal lives, the critics reason.

If we assume this concept to be plausible, we should perhaps ask ourselves just what a Big Brother in the Sky might see. Intimate secrets? Satellite sensors are capable of identifying small farms and large buildings, but hardly have the capacity to penetrate boardrooms or bedrooms. Sensors are, to be sure, already calling attention to fields where farmers are growing crops after being paid government subsidies not to, and pointing out properties and buildings which have been conveniently overlooked in tax records. And they are certainly capable of locating fields of marijuana or heroin-producing poppies, in the United States or abroad. I have found few persons who looked on this kind of synoptic viewing as an invasion of privacy. And, as of this moment, the post-Watergate era seems to be settling into a desire for more open information, not less.

A Big Brother could use a satellite to intercept the telephone conversations carried overland in microwaves by the telephone companies, after which he could, with substantial effort, unscramble them and listen to those he found interesting. (This intercepting capability is not, incidentally, limited to remote-sensing satellites, or for that matter even to satellites. Other equipment could perpetrate such a crime.) But inasmuch as our own government, like others, has access to military satellite information, any Big Brother who would resort to using the civilian NASA remote-sensing information would presumably be an outside-the-government snoop.

Land-based sensors developed by the Army and used in Vietnam have indeed been brought back to the United States and used here in a variety of innovative ways by various levels of government. Audio equipment that once detected troop movements on the Ho Chi Minh trail now listens for smugglers along stretches of the Mexican border. Thermal sensors have been used by U.S. treasury agents in locating stills in several southern states. Low-light-level television cameras have been attached to utility poles in the East and Far West, allowing city police to observe crime (and everything else, its critics complain) in the surrounding streets. A feeling that the Anderson Tapes are being played back in real life has overtaken at least a few scattered critics such as the American Civil Liberties Union.

The rules and regulations governing the use of sensors on the ground obviously must be solved. The same, sometimes agonizing decisions of security versus privacy which have been made in recent years concerning wire taps and computer dossiers in credit bureaus must surely be extended to the world of surface, aerial, and satellite surveillance systems.

As we have already commented, the threat of a satellite sensor with enough resolution to invade personal privacy is by no means present today. But in view of the optimism which Calio and other scientists have about the long-range future of sensing, the problems may not always be academic. By such time as they truly arise, the United States will hopefully have arrived at "ground rules" for land-based sensing, which could then be extended to the skies. The satellite and the sensor, like the computer, may offer dangers and frustrations, but society would seem better advised to learn to control these tools than to stop building them. The maintenance of Spaceship Earth is an essential pursuit for which satellites seem to us an indispensable tool, and such matters as privacy should be studied now, with guidelines determined, before they become urgent problems.

Such matters of concern do extend to the international scene, as demonstrated in a reaction encountered by the University of California's Dr. Robert Colwell on one occasion. Colwell was sitting on a panel at an international symposium overseas when a representative from a country (which Colwell prefers to keep anonymous) sent the panel a note, reading in part, "Our remote-sensing group is enthusiastic about participation in the ERTS project. However, we are not without anxiety about the military and strategic aspects of these new techniques. If we get this free, won't our enemy nations just across the border from us also get free pictures of our country?"[1]

In a sense, the query underlined one of the refreshing aspects of NASA: All the data sensed from every country is available to virtually anyone in the world. Colwell pointed out in his answer that the United States was making available pictures of some sections of the United States that had strategic significance. (He might have added that there had been some hesitation in our own Department of Defense over these disclosures, although the limited resolution has apparently kept Defense from concerning itself unduly.) Each country, of course, makes its decision to apply for ERTS coverage in governmental as well as scientific quarters, before NASA will include a sensing project over any prescribed area. The scientific appeal has been strong enough, and the security exposure apparently minimal enough, that more than thirty countries have participated in ERTS and Skylab projects.

Active Foreign Efforts

Mamadu Konate, as Director of Geology and Mines in the republic of Mali, has his office in an old brick building that was formerly a French barracks. For historical reasons, French was Konate's first non-African language, and Russian, which he learned in order to earn his geology degree in the Soviet Union, his second. He found ample use for it in 1963 when a Soviet geology team began surveying his wild, tropical country and found parts of Mali virtually inaccessible. In their search for minerals, they used the early French aerial photography to narrow down their pursuit on the ground. The pictures, however, proved not to be revealing enough to lead them to minerals, and in 1968 the Soviets finally abandoned what had become a needle-in-the-haystack search.

Konate's major foreign language is English now. He hopes to develop a local survey aimed at high-probability mineral regions. Quartzite shale, a surrogate image of manganese, has been his target on one of the ERTS near-IR bands. If he finds a high-probability zone for manganese, one of the world's minerals in shortest supply, his survey can avoid perhaps 90 per cent of the fruitless areas searched by the Soviets, and have some chance of success. (Thus far, ERTS has yielded one discovery in Mali—not the sought-after quartzite shale, but an unexpected source of water that can be used in agricultural development.)

Mali's success story is one that reoccurred throughout the world in the first months of ERTS sensing. However, most foreign use of data was reduced by 50 per cent in early 1973 when the tape-recording ability of ERTS began malfunctioning. Since all information sensed over the United States could be immediately sent live to one of the three North American receiving stations, our domestic projects were not affected. However, U.S. scientists with Antarctic projects, like most foreign countries, were dependent on the delayed tape transmissions and had their volume of information cut by 50 per cent until the launching of ERTS-2.

Canada, on the other hand, having operated a receiving station since the launching of ERTS, suffered no such inconvenience, and Brazil, in a feat of remarkably good timing, complete a station shortly after the tape failure. The Brazilian governr had, over a period of years, already developed a full-scale operation. Before ERTS, Brazilians had utilized Apollo pictures of their country and, with NASA assistance, had developed a squadron of sensing aircraft. Brazil, with the world's fifth-largest national territory, has been largely unexplored, and stands to benefit tremendously from a combined ERTS-aircraft sensing of her resources. Australia, another immense country with untapped resources, is pursuing an aggressive aircraft-sensing program with sophisticated equipment purchased from U.S. companies.

The Soviets, meanwhile, have operated their own independent remote-sensing program, holding various informational exchanges with NASA scientists with the Soviets usually more inclined to listen than talk. They have operated polar satellites similar to our NOAA series, primarily as weather satellites, but with additional remote-sensing emphasis on polar geology and terrain. Other Soviet missions have been similar to Skylab, except that, so far as we know, the only sensors utilized in that effort have been cameras. "We understand their cameras are multispectral, although the only results shown to us have been in black and white," says William Fischer, a sensing pioneer with USGS and a leading figure in U.S.-Soviet exchange visits.

The Soviet remote-sensing program is an integral part of their military satellite system. "I suspect that the Soviet military has kept tight constraints on the scientists with whom we have met," Fischer has commented.

The greatest area of multilateral international cooperation has been in the area of weather. The World Weather Watch has for many years included exchanges of conventional meteorological information among most of the countries of the world. It has recently been augmented by the Global Atmospheric Research Program (GARP), an even more ambitious effort and one in which space activity has been incorporated. As mentioned, the Soviets have polar weather satellites similar to those of the United States, and they are to be programmed for integrated use alongside our NOAA series, as part of GARP. Japanese, Soviet, and European geosynchronous satellites are to be teamed over the equator with our own Geosynchronous Operational Environmental Satellites (GOES) by 1977. All these interrelated activities are geared to develop the long-range weather-forecasting ability discussed in Chapter 11—a significant goal in a highly computerized interchange of worldwide data.

If we judge by the weather activities, the international sensing effort shows every sign of mushrooming into an immense system. Even considering only the U.S. efforts, we earthlings in our finite spaceship are moving toward examining not only every square mile of potential resources on the surface of our craft, but the ten-mile column of weather above each such square—and, within the next ten years, the stratosphere beyond. What may follow next? Like Marconi, today's scientists are not equipped to answer this question with its universe-embracing implications, although, unlike Marconi and his colleagues, they possess excellent opportunities to communicate continuously among themselves, inadvertently forming a steering committee for Spaceship Earth in the process. Their knowledge, when unwound from individual projects and reconciled into an over-all concept, leads us once again to that simple but exciting conclusion: The big payoff is yet to come.

How Many Dollars?

Few people in the United States, including that share of our populace which is generally aware of vital issues, have any real comprehension of the potentialities (or even of the existence) of earth observations or other useful applications from space efforts. However, these same persons are faithfully informed and reinformed by newsmen of the cost of the space program. With but scant knowledge of the results of spacedom, they are, in effect, repeatedly asked if "moon dust" has really been worth the $40 billion it cost during the big space decade of the 1960s.* For anyone not aware of what the returns on our money may truly come to be, this all sounds immense. Yet that same ten-year ex-

* Total NASA expenditures for the decade, including manned and unmanned programs, space science, and various research efforts.

penditure at NASA is equivalent to the budget for the Department of Defense for a mere eight *months*, or that of the Department of Health, Education and Welfare for a single year. The budget for NASA in 1975 ($3.27 billion) is at much the same level it has maintained thus far in the seventies. In these years, it has been generally equal to the budget of the Civil Service Commission, or of the Labor Department, or about half that of the Transportation Department, or a fourth of that of the Agriculture Department.

At the beginning of the 1970s, the decreased post-moon budgets were severely felt in the space program. In fiscal 1973, NASA was still reeling from a reduction of some 40 per cent from the Apollo peak spending year of 1965. But NASA's chief in the post-Apollo era, Administrator James C. Fletcher, was not looking backward. In December 1972 he wrote that he did not think of 1973 and beyond as "post" anything, but rather as "pre." Space progress, he said, "can best be measured not by the size of our budgets, but by the new knowledge that is being won and by the new operating systems that are being brought into use."[2]

He further defined the 1970s as "a period of intense and realistic preparation for the practical application of space technology in the last decades of this century." In spite of his words of encouragement, Fletcher of course recognizes the limitations of current budgets, as do others at NASA. In particular, scientists who are enthusiastic about Earth Observations sometimes feel that the 5 per cent which the Earth Observations program receives from the entire NASA budget reflects less than a magnanimous spirit. And indeed, if we expect most of the projects we have been discussing in this book to be realized, the scientists no doubt are justified in their feelings. However, a fair appraisal can only be made after noting all the technology which has spun off the manned space program to be enjoyed by Earth Observations. Earth Observations, like private industry and the public, has been handed the technology developed by NASA in its lunar decade; the three main areas of technology in remote sensing—

satellites, instrumentation, and computers—were all, of course, part of Earth Observation's legacy from the Apollo program.

It is our responsibility (and pleasure) as Americans to watch Earth Observations and other NASA projects in the late 1970s, to decide if they merit future funding. Up until now, there has been little effort to relate cost to benefit in the Earth Observations program. We hope, however, that the ERTS-2 project will be evaluated by numerous accountants as well as scientists. In most remote-sensing projects to date, NASA not only has provided the satellites and sensors, but often has even financed the user agency's support work, including ground-truth efforts, related research, and sometimes salaries. But once we pass the Space Shuttle era, and perhaps earlier, we will have arrived at a moment of truth. Users presumably will have to buy their own operational satellites, just as NOAA now pays for its weather satellites. When a user agency begins to pay the bill, rather than piggyback its experiment on NASA-financed programs, we will be more confident that satellites are being used for the proper programs and that significant service is being performed for our tax investment. We look on this adjustment era not so much as a challenge to the remote-sensing concept as a time of defending it from bureaucratic waste and establishing its direction into the areas of most benefit to the earth. This is a step which will have to gradually be taken as part of earth observational transition from experimental to operational age.

A Brain Drain?

While concern about money has been the public's principal objection to the space program, another angry charge occasionally hurled at NASA is that the space program is absorbing too many of the talented brains of the nation. Dr. Wernher von Braun, when he was still with NASA, said he had heard the Earth Observations program "criticized on the basis that it was a shame to siphon off all that wonderful scientific talent from so many different disciplines just to work on the space program."[3]

Dr. von Braun, as we might expect, responded by saying that the reverse was true, that Earth Observations activities provide a peacefully oriented "rallying ground." Certainly the hundreds of scientists in NASA's Earth Observations program are oriented to cooperating on space experiments proposed by the scientific community outside of NASA. The talent that had been "siphoned off" the outside world had come inside NASA to develop the sensing data for use by the universities and governmental agencies. Admittedly the space people sometimes had to promote their technology to outside agencies during the early years, but the requests as well as the funds for future investigations will and should come more and more from the world outside NASA. (If NASA wishes to become a "super-agency," as is sometimes charged, it will have to do so on a subsuper-bankroll.) It should not be the business of NASA to provide relevant projects for the world, but instead to offer the tools, the data, and the expertise for outside scientists to utilize. The demand for this service will have to come from the user agencies and, indirectly, from the American people.

NASA itself would generally agree; the leaders there variously see themselves as explorers, or as scientists offering information which others can adapt to good uses. Even the innovative George Low, who has served the space administration in many top posts and is forceful in promoting his space efforts has stopped well short of suggesting a super-agency role for NASA. What he *does* suggest, as expressed in 1970 when he led NASA as its Acting Administrator, offers an intriguing challenge for the rest of the century. He wrote then, "It is NASA's mission to explore space, to gain a better understanding of the universe, and to use the tools of space to study our earth as a whole. To understand the mechanism of change here on earth, we must understand all of the physical, chemical and biological processes that affect us; we must also understand our history, and the history of the solar system; and we must understand the sources of energy that may be available in the future."[4]

The opportunities and potentialities seem incredibly urgent. The need to control our global spaceship, and the system around it, is fundamental to survival; the earth cannot be isolated. In order to understand our planet, we must, to be sure, continue looking at the earth itself, and yet occasionally we must turn our sensors outward, to inquire into the universe. Low asks, "What physical processes take place in our sun? What are the energy systems of quasars and pulsars? What is the significance of each new finding—for instance, the clouds of formaldehyde and alcohol discovered in interstellar space? "From answers to these questions, and others like them," says Low, "will evolve a blueprint for our environment here on earth.

"And so, while many of today's ills can be attributed to our use of modern technology, the long-range cure—and more important, the prevention of future ills—can only be based upon facts not now well established, and in many cases even unknown." That, believes Low, requires a search for knowledge "in science, and in the application of science, on earth and in space."

Future Research and Exploration

The soundness of Low's statements about "facts not now well established" can best be recognized by noting the effects of previous discoveries as they begin to affect us. We know what has been accomplished to date in our early years of remote sensing, and we have some idea of the immediate potential of the next decade. There are those enthusiasts of agricultural sensing who believe that their present technology can lead to such precise crop analysis that managers will soon be able not only to improve production dramatically but also to distribute their crops to the consumer with remarkable efficiency. There are scientists with similar optimism in virtually every discipline—oceanographers who anticipate that color sensing will unmask most of the mysteries of marine life in the ocean, earthquake scientists who see sensing as the key to meaningful forecasts, hydrologists who anticipate accurately mapping the moisture of the entire globe. Some of the efforts which we have discussed in this book will be realized far beyond expectations, while others will come to unimpressive conclusions. We, of course, cannot predict which shall be which. Even less can we imagine which of those other disciplines which have been barely exposed to remote sensing, and *not* discussed in this book, will reach early fruition; perhaps bacteriologists, who have experimented only briefly in earth observations, will find that sensors can become superbly effective in tracking host environments for communicable diseases. But regardless of which projects achieve success, we can be confident that they will be numerous; we anticipate an abundance of scientific advances through sensing, in this century, which will offer man the potential of mastering a phase of his destiny.

As early as the 1980s man may look back with incredulity on the disorders created today by our failures in anticipating Spaceship Earth's problems—for example, the chaos that occurs in food distribution following an unexpected 10 per cent change in the harvest, or the immediate severity of the 1974 energy crisis that came as a virtual surprise to those persons who had to deal with it, partly because they had possessed no comprehensive and objective information. The matters of food and energy will hopefully be among those benefiting from early progress in remote sensing.

Our lack of ability to foresee which scientific disciplines will develop is evident from the tendency we have seen again and again wherein objectives of a remote-sensing experiment are not met, while success occurs inadvertently in another area. A similar unpredictability has been apparent throughout the history of man and science. Such a pattern has provided the justification (although not the stimulus) for much of man's pure research in all science, including earth observations. Geophysicists and others sought to use satellites to understand the polar regions of the earth for the sake of science, and only later did the broader practical value become evident.

We have spent the preceding chapters discussing remote sensing because it has emerged as the first example of space science affecting the practical needs of earth. But before we close, we need to give passing thought to the immense scope of other space science which has not yet exhibited a practical element. Charles Matthews, as Director of Applications for NASA, looks on remote sensing as the dominant factor in earth applications for this decade—but not necessarily for this century. Within the next few years, sensing may be eclipsed by some other activity with even greater import for the earth.

Earth versus Universe?

Many persons look on our space efforts as a continuing battleground between earth- and universe-oriented scientists. And indeed, there has been a distinct rivalry at NASA between the old guard, with its urges of expeditionary space travel, and the much smaller, newer corps of earth-oriented scientists. But review of the brief history of space science to date would seem to make the choosing up of sides ill-advised. The accidental birth of remote sensing itself is a testimonial to the potential value of continuing forays into space that are not entirely geared to specific goals—much as, in an earlier period, solar science was the eye-opening precursor of nuclear science.

The trips of Mariner spacecraft to Mars have seemed to most laymen to have proved less than they have disproved—reclassifying H. G. Wells's *The War of the Worlds* from a credible drama into fantasy, at least in relation to Mars. (Even this negative discovery, in today's revived era of Unidentified Flying Object controversy, is not necessarily insignificant.) *Positive* developments, however, will surely become apparent in the years ahead. Jupiter, for instance, may yield totally new information useful to man either for his *direct application on that remarkable planet* or as intelligence to be funneled earthward into Matthews' limitless range of projects designed to understand and benefit the world.

Since 1969, the earth has been receiving signals from equipment placed on the moon, developing man's primary knowledge of our own planet's only natural satellite, which space science has only recently determined to be the same age as the earth.

The moon has become our great laboratory. We now know that meteoroids barraged it in the first 600 million years of its life, and simultaneously showered the earth and perhaps the rest of the solar system as well. At the Lunar Science Conference held in Houston in March 1974, such base data was offered to the scientific community for their adaptations to a variety of studies, some of them directly concerning the earth, others the universe.

The useful knowledge which earthlings will possess a generation hence about their own planet and others perhaps may be traceable, in large measure, to sources not yet imagined and achieved with technology yet to be developed. The future of earth cannot be harshly separated from the presence of the universe, nor can earth applications be divorced from the broad-ranging sweep of pure research. The big payoff truly is yet to come, with the possibilities, the benefits, and the excitement which lie ahead so immense as to defy our imaginations.

Notes

1.

[1] D. H. Meadows, D. L. Meadows, Jørgen Randers, and W. W. Behrens, *The Limits of Growth* (Toronto: Potomac Associates, 1972).

2.

[1] Anthony Wolff, "'I love the name,' says Jay Rockefeller, 'I love everything about it,'" *Saturday Review*, August 26, 1972, p. 28.

[2] R. Buckminster Fuller, *Operating Manual for Spaceship Earth* (New York: Simon & Schuster, Inc., 1969).

3.

[1] *Encyclopedia of Aviation and Space Sciences* (Chicago: New Horizons Publishers, Inc., 1969), Vol. I, p. 10.

4.

[1] Earl Clark, "Mt. Rainier—The Live Time Bomb in Seattle's Back Yard," *Science Digest*, August 1970, p. 45.

[2] William G. Melson, "America's Sleeping Volcanoes," *Science Digest*, August 1970, p. 44.

[3] *Earth and Ocean-Physics Applications Program* (draft) (U. S. Geological Survey, May 12, 1972) p. 216.

5.

[1] Meadows *et al.*, *op. cit.*, p. 50.

[2] J. F. Mason and Q. M. Moore, "Petroleum Developments in Middle Eastern Countries, in 1969," *American Association of Petroleum Geologists Bulletin*, Vol. LIV, No. 8, pp. 1524–47.

[3] W. Fischer and E. Lathram, "Concealed Structures in Arctic Alaska Identified on ERTS-1 Image," *Oil & Gas Journal*, May 28, 1973.

[4] Bevan M. French, *Shock Metamorphism of Natural Materials* (Baltimore: Mono Book Corp., 1968), p. 5.

6.

[1] Warren Kurnberg, "Concern for the Arctic Environment," *Saturday Review*, May 16, 1970, p. 486.

7.

[1] *Newsweek*, June 9, 1969, p. 92.

[2] Meadows *et al.*, *op. cit.*

8.

[1] *Remote Sensing* (Washington, D.C.: National Academy of Sciences, 1970), p. 183.

9.

[1] *Water Encyclopedia* (York, Pa.: Maple Press Co., 1970), p. 61.

[2] *Ibid.*

[3] Erwin Cooper, *Aqueduct Empire* (Glendale, Calif.: Arthur H. Clarke Co., 1968).

[4] D. C. Sigmor and V. L. Hanser, *Artificial Ground Water Recharge Through Basins in the Texas High Plains* (Texas Agricultural Experiment Stations, MP-895).

[5] D. Zwick and M. Benstock, *Water Wasteland* (New York: Grossman Publishers, 1971), p. 5.

10.

[1] Georg Borgstrom, *The Hungry Planet* (New York: Macmillan, 1972), p. 379.

[2] Gifford C. Ewing, *The Color of the Ocean* (Washington, D.C.: Earth Survey Office, NASA, 1969), p. 802.

11.

[1] Frederick A. Zito, *Space Program Benefits* (Washington, D.C.: NASA, 1971), pp. 35, 39.

12.

[1] W. Magruder, Joint Economic Committee of the Congress, "Controversy Over the Supersonic Transport," *Congressional Digest*, December 1970, pp. 302, 303.

13.

[1] Robert M. Boyle, *The Water Hustlers* (San Francisco: Sierra Club, 1971), p. 234.

[2] *Ibid.*, p. 232.

14.

[1] Richard H. V. Bell, "A Grazing Ecosystem in the Serengeti," *Scientific American*, July 1971, p. 93.

[2] Robert C. Aldrich, "Space Photos for Land Use and Forestry," *Photogrammetric Engineering*, April 1971, p. 390.

[3] Craig C. Chandler, *Considerations in Detecting and Managing Forest and Rangelands from Spacecraft* (Berkeley, Calif.: University of California, School of Forestry and Conservation, 1972), pp. 9, 10.

[4] *Forecast of Benefits Obtainable by the 1980s from Remote Sensing of Forest and Rangelands from Spacecraft* (Berkeley, Calif.: University of California, School of Forestry and Conservation, 1972), pp. 9, 10.

[5] *Encyclopaedia Britannica*, 1966, Vol. IX, p. 614, and Vol. XVII, p. 631.

15.

[1] *Transcript, Fourth Annual Earth Resources Program Review, Johnson Space Center*, January 18, 1972, Vol. II, p. 485.

[2] *NASA Activities* (Washington, D.C.: NASA, December 1972).

[3] *Transcript, op. cit.*, p. 485.

[4] *NASA Activities* (Washington, D.C.: NASA, December 1970).

Index